Interior Lighting

Created and Designed by the Editorial Staff of Ortho Books

Project Editor
Cheryl Smith

Writer
Kalton C. Lahue

Illustrator
Edith Allgood

Photographer
Kenneth Rice

Photographic Stylist
Ron Boles

Ortho Books

Publisher
Edward A. Evans

Editorial Director
Christine Jordan

Production Director
Ernie S. Tasaki

Managing Editors
Robert J. Beckstrom
Michael D. Smith
Sally W. Smith

System Manager
Linda M. Bouchard

Product Manager
Richard E. Pile, Jr.

Distribution Specialist
Barbara F. Steadham

Operations Assistant
Georgiann Wright

Administrative Assistant
Francine Lorentz-Olson

Technical Consultant
J. A. Crozier, Jr., Ph.D.

Chevron Chemical Company
6001 Bollinger Canyon Road, San Ramon, CA 94583

Acknowledgments

Additional Illustrations by
Ron Hildebrand

Photography Assistant
Melissa McCumiskey

Copy Chief
Melinda E. Levine

Editorial Coordinator
Cass Dempsey

Copyeditor
Irene Elmer

Proofreader
Deborah Bruner

Indexer
Elinor Lindheimer

Editorial Assistant
John Parr

Composition by
Laurie A. Steele

Production by
Studio 165

Separations by
Color Tech Corp.

Lithographed in the USA by
Webcrafters, Inc.

Special Thanks to
Jan Moyer
Nicholas Tuosto

Designers
Carlene Anderson, interior designer, Oakland, Calif., page 60
Agnes Bourne, A.S.I.D., interior designer, San Francisco, pages 3, 56–57
Cynthia Brian, Starstyle Interiors, Moraga, Calif., pages 3, 37
Betsy Corwin, interior designer, Castro Valley, Calif., page 25
Joyce Darrow Interiors, Lafayette, Calif., page 43
The Design Post, Karen Schneble, Danville, Calif., pages 33, 62
Diego Brothers, Inc., builder, San Rafael, Calif., page 25
Pamela Farnsworth, I.S.I.D., Woodside, Calif., pages 3, 10–11, 36, 105
Donna Gleckler & Associates, interior designer, San Francisco, pages 1, 3, 4–5, 15, 22, 25, 32, 41, 46–47
Patty Glickbarg, interior and lighting designer, page 39

Designers (*continued*)
Peggy Grove, Pizazz Interior Design, Lafayette, Calif., page 31
Gwynn-Hogland Interior Design, architect, San Francisco, pages 26, 38, 42, 98–99
T. W. Heyenga, builder, page 14
House of Kitchens, cabinets, Albany, Calif., page 14
Jeff Kerr, Kerr Construction, Nicasio, Calif., page 27
Lighting Studio, Joanne Lim, Berkeley, Calif., pages 20–21, 27, 44
Light Source, Randall Whitehead, San Francisco, page 37
Ben Livingstone-Beneon, neon artist, Austin, Tex., page 48
Luminae Souter Lighting Design, San Francisco, pages 22, 32, 41, 46–47
Casey McGrath Interiors, page 40
Barbara McQueen Interior Design, Walnut Creek, Calif., pages 25, 33
Shelly Masters Studio, faux finish, pages 3, 56–57
Naomi Johnson Miller, Luminae Souter Lighting Design, San Francisco, pages 29, 33, 55
Modular Living, Jock Miller, San Francisco, page 41
Jan Moyer Design, lighting designer, Berkeley, Calif., cover, pages 1, 3, 4–5, 15, 22, 23, 25, 30, 34, 35, 38, 43, 45, 69, 70–71
Yuke Muramoto, builder, pages 3, 56–57
Deborah Rae Interiors, Alamo, Calif., page 30
Rod Rossi, lighting designer, Piedmont, Calif., pages 14, 28, 34, 51, 63
Rick Sambol, Kitchen Consultants, interior designer, Novato, Calif., pages 14, 27, 28, 34, 51, 63
SF 12V Inc., lighting designer, San Francisco, pages 26, 38, 42, 98–99
John Sigg, S & S Construction, Livermore, Calif., pages 14, 29, 34, 51, 63
John Stewart, interior designer, San Francisco, cover, pages 30, 35, 69, 70–71

Designers (*continued*)
Richard Terrell, Terrell Construction, San Francisco, page 37
Nicholas Tuosto, lighting designer, San Francisco, page 25
Chris Volkamer, architect, Point Richmond, Calif., page 24
Homa Yamin, San Francisco, page 18

Front Cover
The interior lighting blends with the fading light of dusk to create a dramatic room.

Title Page
Lighting draws the eye into this home, accentuating art objects and seating areas.

Page 3
Top: Every inch of this kitchen is well lit to make it a safe and pleasant place to cook and eat.
Center left: Hallways can be more than just empty spaces connecting rooms.
Center right: Not all mirrors are lit from the sides. This uplight also works well.
Bottom: This decorating scheme based on fabric and lighting makes for an appealing bedroom.

Back Cover
Top left: A colorful light fixture over this cozy table creates a strong focal point and bathes the nook in a warm glow. A recessed ceiling fixture, hidden from view, illuminates the floral arrangement.
Top right: This well-lighted kitchen features a central ceiling fixture for general lighting and under-the-cabinet lights for the work surfaces.
Bottom left: This classic dining room chandelier makes every meal a grand occasion.
Bottom right: These 12-volt ceiling lamps appear to dance along the parallel wires holding them up. The wires carry—safely and gracefully—low-voltage electrical current, which is converted from standard house current by a transformer hidden in the attic.

Interior Lighting

The Nature of Light and Its Terminology

Lighting Design

Room-by-Room Portfolio

Selecting Light Sources

Selecting Light Fixtures

How to Install Lighting

Maintenance, Repair & Restoration

THE NATURE OF LIGHT AND ITS TERMINOLOGY

Light serves the obvious function of illuminating your home, but it can actually do much more. Well-designed lighting can highlight a treasured possession, improve safety, and even sway emotions. The effect of light depends on a combination of its distribution, its apparent color, and its intensity. Understanding how these aspects of light impact their surroundings is essential to effective lighting design.

Studies conducted in hospitals have shown that patients actually recovered faster when rooms were painted in warm colors and bathed in the optimum amount and tone of light. Elegant restaurants have always paid close attention to lighting for mood, but even fast-food chains now hire lighting designers to improve the atmosphere in their restaurants. You have probably encountered a room that made you feel calmer just by being in it, or perhaps one that seemed pleasant enough but made you want to leave almost immediately. Light can play a large part in generating these reactions.

When designing or redesigning interior lighting, the emotional as well as the visual effect should be considered. Different rooms are meant to evoke different feelings, and lighting can enhance those feelings or destroy them. Understanding the basic concepts of light and lighting will help to ensure that your time and money will be well spent in achieving the effect you desire.

This seating area seems to invite conversation.

HOW LIGHT IS PERCEIVED

One of the less apparent characteristics of light is color. Color is the tone of light as we see it. Most people think of light as being white, but it is really a balanced mixture of different wavelengths that combine to produce the quality called white.

The Color of Light

Electromagnetic radiation is emitted by all objects and reflected by some. A portion of this electromagnetic radiation takes the form of visible waves. These waves are interpreted by the human eye and brain as light. Their intensity is interpreted as brightness and their length as color. When we say that an object is red, we are really saying that the object in question reflects the red wavelengths in light while absorbing the others.

Passing a beam of white light through a prism is a favorite science experiment used to demonstrate the composition of light. The prism refracts or breaks the beam into its different wavelengths. Since the different wavelengths are refracted at different angles, they appear as different colors in a continuous spectrum. Sunlight is composed of all colors of the spectrum: red, orange, yellow, green, blue, and violet. Most people perceive this natural balance as warm, inviting, and comforting.

Electric light is also electromagnetic radiation, but it is created by passing an electrical current through a filament in a bulb or by electrically charging a gas or phosphor coating inside a tube. The color appearance of electric light is determined by the type of light source, which may be incandescent or fluorescent.

Light sources such as the sun, oil or gas lamps, candles, and tungsten filament bulbs all give off incandescent light. The light is called incandescent because it is really a by-product of heat. Electric incandescent light contains most colors of the spectrum, but the majority of the wavelengths are in the red and yellow portion. This warmer quality is not readily noticeable when you turn the light on at night. However, if you turn the light on when there is daylight in the room, the contrast between the two types of light is immediately apparent.

Light given off by standard fluorescent tubes, on the other hand, contains a high percentage of green and blue light waves. This produces a cold, harsh, and even dull or flat effect that can be quite unpleasant in the home. Advances in tube design, however, have produced a wide variety of color-corrected fluorescent lights ranging from the so-called cool tubes to those producing a warm effect. These newer tubes have found a place in the home and can be used without the adverse effects of the older fluorescents.

Color Temperature

Color appearance is described in terms of temperature. Measured in degrees Kelvin (° K), color temperature is usually displayed in the form of a vertical bar or thermometer. Color temperatures at the bottom of

The Electromagnetic Spectrum

Light is only a small portion of the electromagnetic spectrum. All light sources emit considerable infrared energy (heat), and many also give off ultraviolet (UV) radiation.

The Visible Spectrum

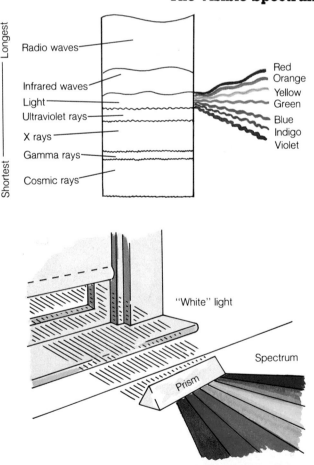

Longest

Radio waves
Infrared waves
Light
Ultraviolet rays
X rays
Gamma rays
Cosmic rays

Shortest

Red
Orange
Yellow
Green
Blue
Indigo
Violet

"White" light

Spectrum

Prism

the thermometer contain a high proportion of red wavelengths and are thus warmer. As the color temperature progresses up the scale, the red is replaced by increasing amounts of blue, and the light more closely approximates daylight.

Color temperature ranges from the yellowish glow of a candle (approximately 1900° K) to direct sunlight entering through a window (approximately 9000° K). The color temperature chart (below) shows where various common light sources fall on the scale and can help you to determine their compatibility in home lighting schemes.

Color Rendition

The perception of light is described in various ways. The term *color rendition* is sometimes used interchangeably with the term *color temperature,* but the two terms actually refer to different things. *Color temperature* describes our perception of light emitted from its source. *Color rendition* describes our perception of the way in which light affects an object it illuminates.

Color rendition does vary according to the color temperature of the light source. White light falling on a green chair shows the true color of the chair. The red and blue waves are absorbed, leaving only the green waves to be reflected.

If the white light is replaced by a green one, there are no red and blue waves to be absorbed, and the color of the chair is intensified. It seems much greener than it really is.

What happens if you focus a red light on the same chair? The red waves are absorbed by the chair, and there are no green waves to be reflected. The chair now seems lifeless, uninviting, and dull.

All of this demonstrates how the apparent colors of fabric, wallcoverings, or paint are affected by the light by which they are seen. Colors that may appear to match each other when viewed by daylight may no longer match when seen under artificial light. Fabrics or other materials seen under the fluorescent lights in a furnishings department can look quite different under the tungsten lighting in your home. It's a good idea to take samples, swatches, or color cards home and compare them with furnishings in your home under your own lighting. By the same token, if you're changing the existing lighting, experiment before you invest in, say, compact fluorescents to see what effect the new lighting will have on your color scheme.

Reflectance

The proportion of light that is returned from a given surface is called reflectance. The brightness of the light striking a surface is measured in footcandles. One footcandle is the illumination falling on a surface 1 foot square from a standard candle located 1 foot away. If 100 footcandles of light strike a wall and 40 footcandles are reflected by the wall, we say that the wall has a reflectance of 40 percent.

In general, the lighter the color, the higher the percentage of light an object will reflect. For example, a white object reflects about 80 percent of the light striking it; a black object reflects less than 5 percent. A white room, which spreads light uniformly and efficiently, has a typical reflectance of about 70 percent. An off-white or cream surface drops the reflectance to about 60 percent. If you repaint walls a darker color or apply darker wallpaper, you will need more light fixtures or higher wattage bulbs just to maintain the same level of illumination.

Surface texture also affects reflectance, although to a lesser degree than color. A light-colored matte finish is the most effective in reflecting light. Thus, a wallpapered room will need more or brighter light than one with painted walls to maintain the same level of illumination.

If the room is large, the surface of the ceiling can be more important than the walls. A ceiling with a light matte finish tends to direct light down into the center of the room, and reduces the level of illumination required.

Color Temperature

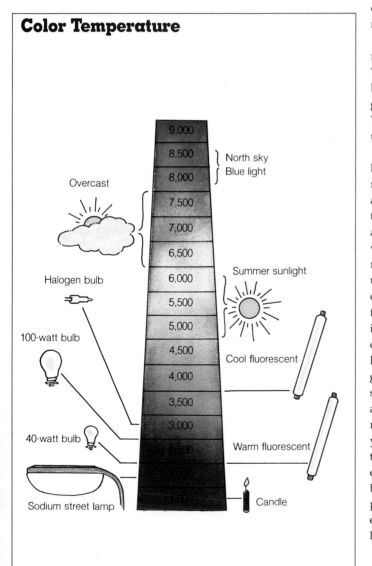

9,000

8,500

8,000 } North sky Blue light

7,500

Overcast

7,000

6,500

6,000 Summer sunlight

Halogen bulb

5,500

5,000

100-watt bulb

4,500

4,000

Cool fluorescent

3,500

3,000

40-watt bulb

Warm fluorescent

Sodium street lamp

Candle

The Distribution of Light

Several factors determine how the light is distributed in a room. The colors and textures of the walls, floor, and ceiling determine reflectance and thus affect the level of illumination required. Consider a small room with a high ceiling. Dark paint can visually lower the ceiling and give the room better apparent proportions, but it can also darken the room. The use of pastel walls and a light-colored carpet can often distribute light adequately enough to offset the darker ceiling.

Masonry surfaces, such as a brick or stone fireplace, occupying a large part of a wall can absorb considerable light. To prevent them from appearing dark, and to assure proper light distribution, such areas can be lighted from an angle. This will increase reflectance somewhat while emphasizing the surface texture.

The types, colors, and textures of the furniture in the room dictate the way in which lighting must be placed in order to maintain an adequate distribution of light. Light-colored furnishings used in a light-colored room tend to increase light distribution. Massive, dark-colored furnishings in the same room will absorb light, reducing its distribution.

Large or numerous windows that allow sunlight into the room during the day and absorb light at night have a varying effect on light distribution. The contrast between sunlight and the shadows it

Color Effects of Interior Lighting

Type of Bulb	Color of Light Emitted	Brightens	Mutes
Incandescent	Yellowish white	Warm colors	Cool colors
Fluorescent			
Cool white	Bluish white	Green, yellow, blue	Red
Warm white	Light amber	Orange, yellow, red, skin tones	Red, blue
Cool white deluxe	White	Virtually all	None
Warm white deluxe	White	Red, orange	Blue, yellow

produces in the room may require the use of artificial light during the day. At night the same windows turn into black holes or cause annoying glare and reflections. It may be necessary to light the area outside the windows enough to balance the indoor lighting, or to provide adjustable window coverings.

Of course, the visual appeal of fixtures is important, but give prime consideration to their functionality. The wrong fixtures, too few fixtures, or improper placement of fixtures can result in an inadequately lighted room, a room that feels uncomfortable, or even an overlighted room.

Contrast

The eye is a marvel, able to accommodate to great variations in lighting—ranging from the less than 1 footcandle provided by the moon to the more than 10,000 footcandles found in daylight. However, a certain length of time is required for the eye to adjust from one light level to another, especially when you are moving from

extremely bright to very dim conditions. When you design a lighting plan for a house, or even for a single room, it is important to take into account how the human eye reacts to differing levels of light.

Home or room lighting should be designed so that the eye does not have to compensate for large, sudden changes in the level of illumination. Plan transitional lighting that will allow the eye to adjust gradually. This reduction in contrast will reduce the shock effect on the eye. It also is safer: A dark hallway between two brightly lit rooms can be not only unappealing but dangerous.

However, a reasonable amount of contrast is desirable. When lighting lacks contrast, the result is a uniform flatness that leads to eye monotony. Differing light levels properly arranged within a room allow the eye to rest briefly as it moves from bright areas to darker ones. Such variations in room lighting also tend to make the room more interesting visually. Properly balanced contrast creates a pleasurable impression. Each room has an

appropriate overall mood; work areas are adequately lit and you gravitate to them naturally; highlights focus attention on attractive features, such as art objects, and shadows obscure less attractive features.

Glare

When a light source is considerably brighter than the rest of the lighting in a room, the eye is instinctively drawn to it and automatically adapts to that level of illumination. Eye specialists call this the phototropic impulse. This phenomenon is encountered whenever there is extreme contrast or when glare is present.

Several forms of glare can result from incorrect or unbalanced lighting in a room. Direct glare caused by a bare light bulb is the most annoying and can temporarily disrupt vision. Veiled glare is caused by a fixture placed directly above a flat, glossy, horizontal surface, such as a table. Reflected glare results from the interaction of a light source and a vertical surface, such as a white, glossy wall. Because glare forces the eye to deal with two very different light levels at the same time, it can cause eyestrain or even a bad headache.

Highly polished or metallic surfaces can reflect the light from a fixture into our eyes. Because light is reflected from an object at the same angle as that at which it strikes the object, an excessive angle of reflectance will produce a "hot spot," or glare. In general, the angle of reflectance from any light source should not exceed 45 degrees.

If a room has uncovered windows, glare may be equally annoying during daylight hours. Sunlight streaming through these windows can cause extreme contrast in one part of the room, making it impossible to see details in adjacent areas. At night, room lights reflecting off the same windows can also cause glare.

The lighting in any room must be flexible enough to provide the proper light level for the activity that takes place in that room. Otherwise you will experience eyestrain and discomfort. Watching TV is a good example. If you watch it with all the lights in the room off, you are straining your eyes, whether you realize it or not. Viewing TV requires a low level of general illumination. Lights must also be correctly placed to prevent them from reflecting off the TV screen into your eyes.

Lighting over a table is best when it is directed at the table top and also produces a soft glow on people's faces. A pendant fixture works well for this, but must be positioned carefully. Reading, doing homework, and playing games are activities that require direct but shadowless lighting, free of direct glare from the light fixture and reflected glare from the book, paper, or work surface.

Of the several ways to combat glare, designing the lighting to prevent it is the most effective. The type and placement of fixtures and furniture should be properly coordinated. Use dimmers to provide maximum control over light levels. Use several low-wattage lamps instead of one or two very bright ones. If downlights are to be used, equip them with baffles or clip-on louvers.

No matter how carefully you plan your lighting, some glare will remain. To eliminate this residual glare, it may be necessary to move either the fixture, a piece of furniture, or a glass-covered picture. Use a shade, if possible, to diffuse the lamp. Change the position of the lamp so that its light falls indirectly on the objects you want to see. Use mats or cloths on dining room tables to minimize reflections. Make sure that lamps are properly positioned when you are reading or working, to avoid both shadows and glare.

Controlling Light Spread

Most people like using table and floor lamps. They come in countless styles, add a decorative touch to one's surroundings, and can be moved for use anywhere in the room. However, people who use these fixtures seldom realize how important the height of the shade and the position of the bulb can be.

Using the correct shade on a table or floor lamp is vital for good light distribution. Because a shade directs the spread of light given off by the bulb, you must use one that will send the light where you need it.

Light spread increases as the bottom diameter of the shade increases. Too narrow a shade can keep the light from reaching the entire area you want to illuminate. A shade that is too wide can spread the light too far. An adjustment of as little as 2 inches in the diameter of the shade can make a noticeable difference in the light spread.

The same problem will arise if you replace a bulb with one that is physically larger or install a touch control or other adapter between the bulb and the socket. In each case the filament of the bulb will be positioned higher in the shade than before, reducing the light spread from the bottom of the shade. This problem can be corrected by using an extension screw on the lamp harp to raise the height of the shade, or by changing the harp itself.

Light Spread Controlled by Shades

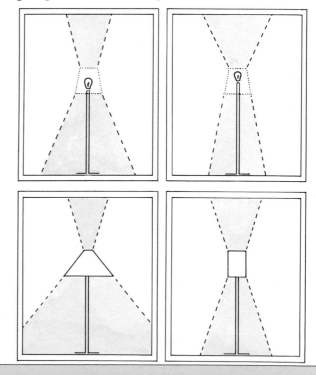

LIGHTING DESIGN

Good lighting is effective lighting. It can control the flow of people through a room, indicate where they sit or work, and make them feel comfortable and at home. Good lighting does not stand out as such. When you enter a room with a good lighting design, your eyes tell you that the entire room and everything of importance in it can be seen with ease. Mentally you accept this as the norm and so pay little attention to it. Poor lighting, on the other hand, stands out immediately. Your eyes quickly tell you that something is wrong, and you start to see things that would otherwise escape your notice.

Successful lighting design begins with a carefully thought-out plan. In this section you'll find out why you need a lighting plan, how to assemble one, and how to get professional advice when necessary.

Well-designed bedroom lighting often has separate controls for each side of the bed, allowing one person to sleep while the other reads.

DESIGNING LIGHTING

Whether you want to replace or simply to improve the lighting in your entire home (or in just one room), you will need a lighting plan. A plan is even more critical when you are building an addition to the house. This is the time when it is easiest to avoid mistakes and least expensive to install proper lighting.

The Lighting Plan

Many people stumble into changing their home lighting because they find an attractive fixture in a lighting shop. They buy the fixture and take it home with the idea that they can fit it into a room, only to find that it clashes instead of blending, or that it doesn't deliver the kind of light they expected. Obviously, this approach to lighting design should be avoided. Selecting a light fixture is one of the last things involved in implementing a new lighting scheme; an attractive fixture cannot compensate for bad lighting.

The starting point for any lighting scheme involves the room itself and the activities for which the room is used. Once these are defined, you can devise an appropriate strategy for lighting the room. Because lighting relates directly to room size, decor, and furniture placement, it is as important to the decorating scheme as are color and style.

The amount of work involved in designing a lighting plan depends on the size of the project. You don't need a complicated plan just to rearrange a few pieces of furniture. However, if you are remodeling or building an addition, the lighting plan should be more expansive, and it should include a room sketch. Whether it is simple or complicated, the primary purpose of any lighting plan is to make certain that important details, such as installing enough outlets for plug-in task lighting, are not overlooked.

You need to consider such things as the shape and size of the room, the placement of furnishings, and the type of activities for which the room is used. When dealing with your entire house or apartment, relate the rooms to one another, as well as to adjacent connecting areas, such as halls or stairways, to make sure that there is proper transitional lighting. A floor plan sketch is especially useful for this type of planning. If you're adding a room to the house, you can base the floor plan sketch on the architect's blueprints or plans.

This sketch has several uses. First, it can help you to decide where to position the most important pieces of furniture. Once you have decided where to place the furniture, it's far easier to determine what kinds of fixtures you need and exactly where they should be installed. Second, the sketch lets you see the geometry of the space more easily. Built-in lights are generally arranged in a simple geometric pattern, but they should also relate to the activities and furniture groupings of the room.

Next, the sketch can help you to decide how many outlets and switches are required and where they should be located to provide the most convenient control system. (In an existing home, it is also a good idea to walk through the rooms to determine the best location for each switch.)

Finally, the floor plan sketch will help you to estimate the cost of the project, whether you price it out yourself or decide to get estimates from professional electricians.

A lighting plan is a problem-solving tool. It helps you to determine where light is wanted and where it is needed. Once you have a plan, you can concentrate on implementing it inexpensively and with style.

Lighting Designers and Consultants

If you are not confident of your own ability to design good lighting, you can get professional advice for a reasonable fee. Ask friends for recommendations or write to the International Association of Lighting Designers (30 West 22nd Street, New York, NY 10010) for the names of members in your area. Always ask to see a lighting designer's portfolio.

Lighting designers offer various levels of service. They can advise you how different types of lighting might be used in your home, and they can solve lighting problems in ways that you probably wouldn't think of, because they are up-to-date on both trends and fixtures. They can design a specific lighting plan and suggest specialized or custom-made fixtures to meet your needs. A few designers will even supervise the installation of the lighting plan to make sure that the work is done properly and meets all code requirements.

You might retain a designer or consultant simply to provide a series of proposals that you or an electrician can use to do the work. In many cases this is more economical than letting the designer do the entire job. It can also be more economical than trying to do the job yourself, since the designer will often suggest cost-saving measures that you might otherwise overlook.

Cost Estimates

Before you try to estimate the cost of implementing the lighting plan, you must have a clear idea of how many fixtures, receptacles, and switches you will need. Do some shopping to determine the approximate cost of the hardware, and, if you are planning on using them, contact at least three licensed electricians. Give them each a set of exact specifications—what you want installed and where—and ask for detailed estimates on the cost of installation. Make sure that the estimates state exactly what work will be done as well as the length of time required to do it.

Add the estimated installation cost to the price of the hardware and compare the total figure to your budget for the project. If the project will cost more than you can afford to spend, you may decide to take a second look at the lighting plan, or to implement it in stages instead of all at once.

The Lighting Plan

Floor Plan Empty

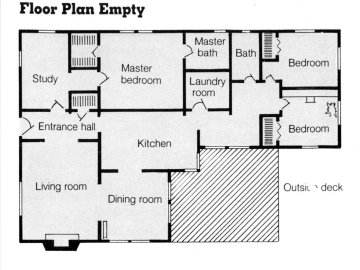

Study

Master bedroom

Master bath

Bath

Bedroom

Laundry room

Entrance hall

Kitchen

Bedroom

Living room

Dining room

Outside deck

Floor Plan With Furniture

Floor Plan With Furniture and Lights

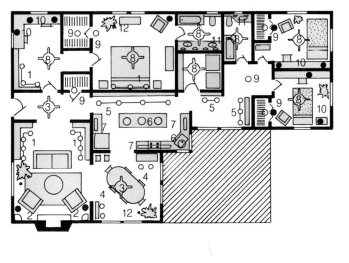

1. Track lights to light pictures, stereo equipment, books, spotlight plant, for reading

2. Wall mounted, indirect lighting

3. Chandeliers

4. Recessed downlights to light dish and linen

5. Track lights to light books, pictures, knickknacks

6. Pendants light work surfaces

7. Undercounter lighting

8. Strong overhead lighting

9. Recessed downlights (closets, dressing area before mirror)

10. Desktop lamps

11. Around mirrors

12. Spotlighting for exotic plants

13

If you find it necessary to economize, or if you feel that you must scale the project back to fit it into your budget, the best way to proceed is to install the wiring all at once and to install fewer or less expensive fixtures. Installing wiring in stages involves a major disruption and mess at each stage, and it actually ends up costing more. Although the less expensive fixtures may not be as visually attractive as the ones you had in mind, they can easily be upgraded later, when money is less of a problem. Of course, since recessed fixtures are not seen and are relatively inexpensive, but the cost of installation is high, all these fixtures should be installed at once.

Types of Lighting

Lighting can be separated into three basic types or categories: general (ambient) lighting, task (work) lighting, and accent (mood or display) lighting. A good lighting plan makes use of all three types to light an area or room according to its function and style. The distinguishing characteristics of each type are discussed below, and the most appropriate fixtures are considered briefly.

General Lighting

General, or ambient, lighting replaces sunlight, filling a room with overall illumination. Its primary purpose is to let you see your surroundings and to enable you to move safely around the room.

If your general lighting is indirect, much of the illumination it provides results from light that bounces off the walls, ceiling, and floor. To make indirect light most effective, these surfaces must have a high level of reflectance. Since pale walls and ceilings reflect more light than dark ones, they are more efficient in creating a soft and uniform level of illumination. Matte finishes are preferred, offering a high level of reflectance without creating the glare and highly visible surface imperfections of a gloss or semi-gloss finish.

Reflectance also affects the apparent size of a room. The walls do much of this work in small rooms, and the ceiling and floor do it in large rooms.

Fixtures such as uplights, torchères, and some wall lamps are specifically designed to provide indirect light; pendants, table lamps, and floor lamps radiate light toward both the ceiling and the floor. Indirect light can also be produced by recessed ceiling fixtures with diffusion panels, or by concealed light fixtures hidden behind cornices, valances, and wall brackets.

Task Lighting

Specific activities performed in a room require a smaller but much brighter area of illumination than the general lighting can provide. The purpose of this focused light is to permit a visual task to be performed without discomfort or eyestrain.

Good task lighting must do more than simply provide brighter illumination. The light must be properly positioned to avoid causing glare or casting unwanted shadows.

Not all tasks require the same amount or type of light. How much light is necessary depends upon the task and the amount of detail that must be seen with ease. Cleaning a kitchen appliance does not require as much light as sewing or needlework. Ironing, however, is made easier by an oblique light that reveals wrinkles and creases. Unlike most other tasks, painting requires light with specific color-rendering properties.

Avoid pitfalls such as overlighting and excessive contrast when you choose task lighting sources. Casual reading requires somewhat less light than similar activities because it is assisted by the contrast between the page and the print. Yet prolonged reading or reading by an older person can require up to twice the light needed for casual reading. Contrast between the work surface and the surrounding area should be balanced to prevent eyestrain. The rule of thumb used by experts is to keep the surrounding area at least one third as bright as the work surface.

Task lighting can be provided by recessed or track fixtures, pendants, table or floor lamps, and undercabinet

Recessed fixtures provide the diffuse light that fills this kitchen.

lighting. Tungsten lamps are generally used for task lighting, but avoid high-wattage bulbs; they can create unwanted heat. For hobby activities such as mounting stamps in an album, a halogen lamp provides the high level of illumination necessary without giving off heat that might damage the stamps. Tubular fluorescent fixtures are good for lighting a kitchen counter or a workshop bench, whereas circular and other small fluorescents work well for smaller areas. When lighting long areas such as kitchen counters with fluorescents, the accepted rule of thumb is to provide 8 watts for each foot of counter length.

Accent Lighting

Lighting can also be used in various ways to set a mood or add drama to a room or area. Primarily decorative and most often directional in nature, accent lighting can be arranged to spotlight paintings, sculpture, or collectibles; to draw attention to architectural features; or to highlight the texture of a wall, draped fabric, or masonry. Regardless of what you are accenting, it must be lighted properly if you want to show it to its best advantage.

Accenting requires a light source that will throw at least three times more light on the object or area to be accented than the general lighting in adjacent areas. Proper positioning of the light is as important as its intensity. As a rule of thumb, designers recommend that the fixtures be aimed at a 30-degree angle from the vertical to prevent their light from interfering with the line of sight, and to avoid casting disturbing reflections on the surface of an object.

Recessed downlights or track lights are good choices for homeowners; renters are more inclined to use track lighting. Track lights have the advantage of letting you move the fixtures to different positions as required by the objects you are accenting. In general, whether you choose track lights or downlights, you should use one fixture for each object being accented. With three-dimensional art objects, such as sculptures or vases, use two or more light sources aimed at different angles to create a more desirable accent.

Directional and primarily decorative, accent lighting can be used in many different ways. Experts recommend, however, that spotlights be aimed at a 30-degree angle for best results.

FUNCTIONAL LIGHTING

The first rule of functional lighting is to put light where it is needed. However, simply doing this does not guarantee that the lighting will be attractive, comfortable, or even appropriate.

Achieving Proper Balance

Lighting that works only from a functional point of view is often unsatisfactory and even uncomfortable. You have only to imagine old-style government offices with their row after row of desks beneath banks of recessed fluorescent fixtures to see the results of functional lighting that lacks proper balance.

In this section you'll discover how to achieve balance by structuring and combining light. Once you have the concept in mind, you'll be able to apply structuring to your particular needs and appreciate what a difference it makes in the final results.

Structuring Light

An effective combination of bright, medium, and dim light provides visual interest and contrast as the eye moves around the room. Varying the amount and intensity of light according to how the room is being used at the time results in a space that is both friendly and pleasantly lighted. Structuring the three levels of light is the primary technique for obtaining proper balance in room illumination.

The first step in structuring the lighting of a room is to define its primary focal points. These are the areas where

major activities take place. Designers recommend the use of at least two focal points—three if possible. The brightest lights (Level 1) should be positioned to illuminate these areas. Level 1 light is task lighting; it can take the form of table lamps, floor lamps, downlights, or track lights.

The second, or medium, light level fills in the gaps between the focal points. Level 2 lighting should be bright enough to connect the focal points and to provide both contrast and continuity in vision when one looks around the room, but it should not be bright enough to be distracting. It can be either direct or indirect light provided by pendants, uplights, or any of the fixtures used for Level 1 lighting.

The third, or dim, light level is necessary to fill in dark areas and to define the boundaries of the room. It generally takes the form of indirect or diffuse lighting.

In general, the ratio between Level 1 and Level 3 lighting (whether measured in lumens or in footcandles) should be about 3 to 1 and no greater than 4 to 1. Ratios greater than 4 to 1 can certainly be dramatic, but they are not satisfactory for normal activities.

Combining Levels of Lighting

Generalizations about combining the different types of lighting should be approached with care. Although the theory behind such generalizations is sound, it is not always practical, or in some cases possible, to apply them to a particular room. In many living rooms, for example, general lighting may come not from one or two primary fixtures but from a combination of task or accent fixtures.

Experts agree that there is no single correct way to light a room, nor is there an optimum light level. Although concerns about space, furnishings, and decor play an important role, mood and your own taste are equally important considerations. The purpose of combining different types of light in one room is to create different light levels that will help the room to work as intended and will give the eye an opportunity to play in the space without strain.

Measuring Light

It is difficult to define a particular level of illumination as "bright," "medium," or "dim." There are, however, ways to determine a suitable level of brightness and what is required to produce that level. To do this intelligently you should know something about how light is measured.

People differ in what they perceive as bright, medium, or dim light. Your own feelings about this will change according to the time of day, whether you are indoors or out, the relative brightness or darkness of

the surroundings, and the type of lighting to which you are accustomed. However, it is helpful to establish some standards for levels of illumination in terms of footcandles, as well as to determine the output of a light source in terms of lumens.

Footcandles

A footcandle is the unit of measurement that defines the brightness of light falling on a surface. One footcandle is the illumination striking 1 square foot of surface from a standard candle located 1 foot away.

For our purposes we can define dim light as light of 4 to 7 footcandles. This level of illumination approximates the necessary ambient lighting in a living room or hallway. It is sufficient to permit safe movement through the house and comfortable TV viewing.

Medium light has a brightness of 8 to 25 footcandles. Many indoor environments are lit to this level of illumination, which is sufficient to perform general activities that do not require seeing great detail for any extended period of time. These would include dining, conversation, casual reading, and playing with the children.

Anything above 25 footcandles can be considered bright light. At the lower range this is task lighting; a small reading lamp illuminates a book to about 50 footcandles. Food preparation and hobbies, such as stamp collecting and needlework are other examples of activities that require this level of illumination. Levels of 75 to 100 footcandles constitute accent lighting, which is used to spotlight artwork or to emphasize architectural features.

Lumens

A lumen is the unit of measurement that defines the amount of light emitted by a lamp. One lumen is the amount of visible light energy emitted by one standard candle at a distance of 1 foot.

While a footcandle is the amount of illumination at a given point (1 foot), a lumen is the amount of light energy reaching that same point. For all practical purposes, 1 lumen falling on 1 square foot of a flat surface produces an illumination of 1 footcandle.

Lumens are measured by the manufacturer of the lamp. The manufacturer prints the average lumens given off by a particular type of bulb on the sleeve or box in which the bulb is packaged. This average applies only to new bulbs. Bulbs can lose as much as 20 percent of their lumen output over their life.

In planning light levels you can add up the total number of lumens emitted by all the light sources in a particular area or room. Lighting designers recommend between 1500 and 2000 lumens for watching TV. Precise work such as needlepoint, which involves both accuracy and contrasting colors, requires a minimum of 2500 lumens, a large percentage of which should be directed where the work is being done. Of course, these figures in themselves mean little. The required number of lumens is greatly affected by the size and shape of the room being lit.

Fixtures that use three-way bulbs are helpful in adjusting light output according to the desired tasks. Several small bulbs in fixtures near the work area can provide the total lumen output required for work activities, but using the concentrated light output of one or two bright bulbs is more cost-effective. An alternative to a three-way bulb is a screw-in device which allows you to change any lamp socket into a dimmer, thus allowing you to choose the level of light appropriate for the moment.

Determining Light Levels

From the dawn of the electric age, Americans have enjoyed bright light and plenty of it. Whether because lighting is relatively inexpensive in the United States or because American lighting engineers have consistently recommended higher levels than their counterparts abroad, the fact remains that ample indoor illumination is a part of the way of life in this country.

There is no handy rule of thumb to use in determining appropriate light levels. In general, people choose a level of illumination that meets their individual needs and tastes. Some people who work in brightly lighted environments grow accustomed to the high level of light and carry it over into their home life, while others retreat into more relaxing and lower light levels when at home.

There are, however, certain factors that must be taken into account when determining a satisfactory level of illumination. First, consider the activity you are performing. Different activities require different amounts of light, according to the detail involved in the activity. For example, lighting experts recommend twice as many footcandles for preparing food as for casual reading. The length of time to be spent at a task also impacts the amount of footcandles required. More light is needed to perform a given task for several hours than to perform the same task for only a few minutes. Finally, consider the age and eyesight of the persons involved. A fifty-year-old may require up to twice as much light as a twenty-five-year-old to perform the same task. Some provision should be made to accommodate the needs of both people.

How to Measure Footcandles

Lighting engineers and designers measure footcandles with expensive specialized meters, but you can do it at home with any 35mm camera equipped with a built-in light meter. Photo hobbyists who have an incident light meter will find that it too works very well, since this type of hand-held exposure meter accurately measures the light falling on it and reads directly in footcandles.

To use a 35mm camera, set the film speed (ASA or ISO index) dial to 100. Position a large sheet of white paper or posterboard at a 45-degree angle on whatever plane you want to measure. For example, if you want to measure the light falling on a tabletop or countertop, prop the paper or posterboard up on the surface at a 45-degree angle.

Point the camera at the paper, taking care not to cast any shadows on it. Fill the viewfinder frame with the paper. If the viewfinder sees more than the paper, the meter reading will be incorrect. Now adjust the camera to provide an exposure at f/4.

Note the shutter speed required for the exposure and consider it as a whole number instead of a fraction. This will tell you the approximate number of footcandles falling on the surface. For example, if the shutter speed is $1/60$, the light reaching the surface is approximately 60 footcandles; if the speed is $1/100$, the light is about 100 footcandles.

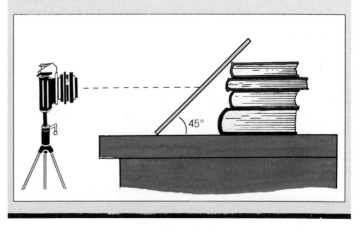

Even if you have devised a complete lighting plan to meet your current or anticipated needs, there may be good reasons why you cannot put it into practice. There are, however, plenty of ways to improve lighting, whatever your circumstances.

Meeting Lighting Needs

If you rent your house or apartment, your lease agreement may limit the alterations you can make to the property. Homeowners are not subject to such prohibitions, but the cost of redoing your lighting scheme as planned can be prohibitive. Even homeowners should consider how the future value of their homes may be affected by radical designs.

Buyers of newly constructed houses often find a variety of improvements desirable, since builders seldom pay much attention to lighting beyond installing a minimum number of fixtures and receptacles. Unless the house is custom-built, builders generally leave it up to the home buyers to design and install their own lighting arrangements, according to their needs, tastes, and budget. Failing to recognize this, many new buyers often feel a vague dissatisfaction with the lighting in their new homes without realizing why. Even those who understand the problem may not be able to stretch their budget to accommodate a complete revision of their lighting.

If any of these factors interferes with your lighting plans, there are various inexpensive ways to upgrade or improve your present lighting to make it more suitable without knocking out walls or rewiring a substantial part of your home.

How to Improve Present Lighting

If you want to make your present lighting more effective but cannot invest a great deal of money at this time or make structural alterations to the property, there are a number of commonsense approaches.

One of the first and easiest things to do is to inventory all the bulbs in the fixtures. See whether they are doing their job or whether another type of bulb might be more suitable. For example, a bulb that is too bright may cause unpleasant glare; one that is not bright enough may cause gloomy spots in room lighting.

Check downlights to make sure that the bulbs are of the proper type and wattage. The previous tenant or owner may have installed regular tungsten bulbs in place of the reflector bulbs that these fixtures were designed to use.

Take a close look at any fixtures fitted with a shade. Replacing an opaque shade with one that diffuses light can open up areas in the room. By the same token, using an opaque shade can reduce the level of illumination provided by the fixture.

Dimmers are easy to install and give you considerable control over fixed or plug-in lighting. Three-way bulbs offer another form of control. Clip-on fixtures operate from wall outlets and can be used for task or accent lighting. Adjustable task lights, such as a desk lamp, can be used to bounce light off the wall or ceiling when they are not being used for their intended purpose.

Track lighting and surface-mounted raceways or plug-in strips offer renters a less expensive means of increasing their options. These fixtures can be easily installed and, when moving day arrives, can be removed with minimum surface damage. Don't think of track lights as spotlights only. Remember, there are adapters that permit the use of pendants or chandeliers at any point on the track (as well as suspension hooks for hanging plants). If you use track lighting with a plug-in connector, take pains to hide the cord as much as possible. How this is done depends on the room and the position of the track.

Fixtures can be attractive as well as functional, as this bathroom illustrates.

New Lighting

You'll never have a better time to design and implement a lighting plan exactly as you want it than during new construction. Whether you're custom-building a house, putting an addition onto an existing structure, or shelling a room and starting over, you have the perfect opportunity to make the interior lighting an integral part of the overall scheme. By designing and coordinating the lighting as you work out the details of the interior design, you can select colors, finishes, light fixtures, and light sources that work together in harmony.

This is also the time for long-range planning; the time when you have the best opportunity to anticipate and provide for possible alternative uses of the space. For example, it might be necessary or desirable in the future to convert a study into an extra bedroom. Certainly the lighting requirements in a child's bedroom will expand and change as the child grows from toddler to teenager.

The number and location of plug-in receptacles and switches is one of the most important, yet often overlooked, considerations, since the lack of either can spoil an otherwise excellent lighting scheme. Plug-in receptacles add to the versatility of a room, and you can rarely have too many of them. The same holds true for switches, which should be located along the main circulation routes in the house.

Another oversight with new construction is the failure to include daylight in the lighting plan. The lighting requirements in a room used primarily during daylight hours will differ from those in a room used mainly after dark. Not only does this have a bearing on the colors and finishes used in decorating the room, but it also relates to the number of fixtures required and the wattage of the bulbs used in those fixtures.

The best way to avoid overlooking the obvious is to make a wiring plan on paper. This is similar to the floor plan sketch described on pages 12 and 13.

Using graph paper, carefully draw the outlines of each room on a scale of 1 inch to 4 feet. Include windows, doors, and any built-in features, such as bookcases or shelving. Make several copies of this room outline so that you will be able to compare various arrangements.

Plot the major pieces of furniture on the outline first. Then place the permanent light fixtures at appropriate locations. You'll never have a better opportunity to plan the installation of downlights or other recessed fittings without worrying whether there's enough space to accommodate them.

Indicate the various spots where plug-in lighting would be desirable. You need not actually install portable lighting in all these places, but if you know the desirable locations, you can give special attention to the placement of plug-in receptacles, switches, and dimmers. This ensures that your control system will be adequate to handle several variations in lighting should a future need arise.

Swag Lighting

Ceiling fixtures suspended from a chain are far more versatile than they might appear at first sight. This design lets you adjust the height of a light simply by adding links to, or removing them from, the chain. The chain carries the weight of the fixture, camouflages the cord, and can be used to swag the fixture to position it at any point on the ceiling.

This last feature can be invaluable when it comes to redecorating or even remodeling. For example, suppose that your dining room table is centered under a rigid ceiling fixture—an arrangement common to many dining rooms. When the new table you've bought arrives, you decide that it looks better next to the picture window than it does in the center of the room. The rigid ceiling fixture, however, remains centered in the room, where it no longer serves its purpose. With a chain lamp instead of a rigid fixture, you need only a couple of hooks and a length of chain to reposition the light where it can do its job. Some rigid fixtures can be swagged by using a swag lamp conversion kit.

Chain lamps have other advantages. They can take the place of table lamps in homes where small children might knock a table lamp over. They can be positioned wherever you want them, even if you have no ceiling outlet, or if the ceiling outlet isn't located where you want the lamp.

It is not difficult to swag a chain lamp. If the room has no ceiling outlet, you can extend the wiring and chain far enough to plug the lamp into a wall socket. Install a ceiling hook where you want to hang the lamp and run the chain and wiring up to the hook. If you have a ceiling outlet but it is not located where you want to drop the lamp, extend the chain and wiring as required. Then swag the chain from the outlet to the desired location and hang the lamp with ceiling hooks.

You will need a pair of chain pliers to open and close the links, as well as one or two chain hooks. Hooks come with screws for use in plaster or wood and toggles for use in wallboard. One hook should be sufficient to swag a lamp from a ceiling outlet. If you're running the lamp more than 4 feet from the ceiling outlet, or from a wall socket to the ceiling and then across, you'll need two hooks.

Install the first hook directly over the spot that you want to light. Then decide how much drape or swag across the ceiling looks best and install a second hook if necessary. If the light is being swagged from a wall outlet, the second hook can be located either on the ceiling near the wall, or on the wall near the ceiling. When you hang the lamp, only the chain link should be slipped over the hook. Let the chain carry the weight of the fixture to avoid any drag on the cord that might stress the connections.

ROOM-BY-ROOM PORTFOLIO

Different areas of the home require different lighting. To assist in your planning, here are a variety of design solutions for rooms as well as transitional spaces. Use them to provide ideas for your own lighting needs. Look at how various levels of light create different feelings in the rooms, how placement of fixtures can help in defining areas within rooms, and how collectibles and paintings become even more attractive when properly lit.

This dining room combines drama and usefulness, with its carefully positioned pendant, accent lighting, and view of the city lights.

ENTRANCES

Entrance lighting welcomes visitors and helps to establish the overall atmosphere of the home. For this reason it should be warm and inviting, with enough illumination to help the eye make the transition from outdoors into the living area.

Don't overlook the outside lighting at the entrance. Its brightness plays an important role in determining the entrance lighting. If the lighting outside is dim, the entrance lighting should be subdued in order to accommodate the eye and allow adjustment to brighter interior lighting.

The size of the entrance area determines both how it is used and how it is lighted. Small entrances used primarily for storing outdoor clothing, umbrellas, and footgear are limited to this function because they lead almost directly into the main living areas. Indirect lighting from wall sconces or a circular fluorescent in a diffusing ceiling fixture can provide enough illumination to make the visual transition from outdoors to indoors, and to ensure safe passage into the living areas.

Large entrances have a wider range of functions. When they are used to exhibit pictures, hold furniture, or display plants, the lighting can be more dramatic, since most people spend little time in an entrance.

A long entrance that turns into a hallway with rooms off one or both sides can be variously lighted. One approach is to lead the eye into the living area with fairly strong directional lighting from the ceiling or the upper part of the walls. Another is to recess fluorescent fixtures in the walls, with baffles to direct the illumination upward. The light will bounce off the ceiling for an open, cheery look.

Uplights are ideal in large entrances, but they should be used only when there is sufficient space to arrange the fixtures without posing a safety hazard. Closets and mirrors in the entrance area should have their own lighting, making it easy to find one's coat or to check one's appearance.

The American Lighting Association offers recommendations for entrance lighting. With incandescent bulbs use 100 to 120 watts for areas of less than 75 square feet, and a minimum of 150 to 180 watts for larger areas. With fluorescents use 26 to 40 watts for small areas and a minimum of 30 watts for large ones.

The fine lighting in this entrance not only draws visitors into the home, but also highlights various art objects and a useful hallway mirror.

This grouping is the focus of the entry, washed with light.

Large entries offer many possibilities. Subdued illumination allows for the dramatic highlighting of art and draws the eye to the brighter rooms beyond.

KITCHENS

The kitchen is usually the first place where the family congregates in the morning, and often the last place where they meet at night. It serves as a gathering place for teenagers and is used for entertaining; it is often where household management tasks, such as the family bookkeeping, are done. Appropriate lighting is especially important.

Since the kitchen is used extensively both during the day and after dark, it is important to consider the natural lighting. Well-designed kitchens will have plenty of daylight illumination; they will require little in the way of supplemental lighting until dusk. Kitchens with insufficient natural lighting should be provided with useful artificial illumination for daytime as well as nighttime use of the room.

In both instances good general lighting will be needed at night and on dark days. The light level should be high enough to provide easy visibility inside cupboards and cabinets. You can drop the ceiling and install recessed fixtures if you are remodeling an old house. The kitchen is the area most often chosen for remodeling, incidentally. Real estate studies confirm that the money is well spent; the cost of a well-lighted kitchen is often recovered completely when the house is sold.

Install strong, shadowless task lighting at strategic points: over the stove, above the sink, and over the countertops. The kitchen poses numerous safety hazards; good task lighting will help you to avoid cuts, burns, and scalds. The fixtures should be efficient in providing light and should be easy to clean.

The low operational cost of fluorescents has long made them popular for use in kitchens, but tungsten lighting provided by pendants has returned to favor. Many modern kitchens now combine the two sources, using general tungsten lighting from pendants and task lighting from fluorescent strips hidden under wall cabinets. To prevent countertop glare, mount the strips as close to the front of the cabinets as possible and make sure that the fixture covers at least two thirds of the counter it illuminates. The rule of thumb is to provide 8 watts for every foot of counter length.

Kitchens can combine nearly every sort of light fixture. The design should include both general and task lighting.

Although the lighting systems in these kitchens vary widely, all combine general and task lighting to ensure that cooking activities are well illuminated. Fluorescents, now required by some local building codes, can be blended with task lighting for a utilitarian and attractive kitchen, although incandescents are available in a wider variety of fixtures.

This unusual 12-volt lighting system suspended below the wood ceiling combines well with task lighting to produce a workable kitchen.

BATHROOMS

Safety is critical to bathroom lighting. Because water and electricity can be a lethal combination, fixtures and bulbs must be positioned so that they will not be splashed with water or touched with wet hands. Shower lights should have a wet-location recessed fixture or a surface-mounted wet-location fixture.

Large, multiuse bathrooms have become fashionable, with their emphasis on relaxation and fitness as well as on grooming. Today's bathrooms can be fitted with such amenities as exercise equipment, whirlpool tubs, and sauna or steam compartments. The larger and more versatile the bathroom, the more attention must be paid to lighting that is both decorative and effective. Large bathrooms can benefit from carefully planned mirror installations, which spread light very efficiently while creating the feeling of open space.

Install low-voltage minilights between the mirror panels to add a touch of sparkle and create an unusual night-light.

A central ceiling fixture provides the overall lighting in many bathrooms. This fixture may use a tungsten bulb enclosed in a diffuser, or it may use a circular fluorescent bulb. Although this may sound unimaginative, it is often the best approach in small bathrooms. Large rooms can make good use of structural lighting, such as a dropped soffit installed above a vanity. General illumination requirements range from 100 to 150 watts for incandescent lamps and from 60 to 80 watts for fluorescents.

The real key to good bathroom illumination, however, lies in designing even, shadowless task lighting. Bath-vanity strip lighting is a popular choice; it provides excellent task lighting for shaving, applying makeup, or general grooming. Install strips along both sides of the mirror, with 15-watt to 25-watt white G (globe-shaped) bulbs. This very practical form of lighting adds a touch of glamour to the bathroom, and the warm colored light from the tungsten bulbs flatters skin tones.

With the newer color-corrected tubes, fluorescents can be used successfully at the bathroom mirror.

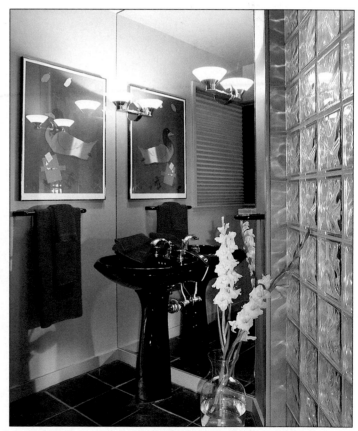

The small spaces in most bathrooms can often be lit by one well-placed fixture, but several fixtures may produce a more pleasing design.

LIVING ROOMS AND FAMILY ROOMS

The main center of activity in a house or apartment is the living room. It requires a highly flexible lighting system. Whether the fixtures are used alone or in combination, they must provide a sufficient quantity and quality of light to accommodate activities and to create a variety of moods for entertaining or for spending a quiet evening at home.

Because personal tastes play a large role in living room lighting, there is no expert consensus as to what is best, but experts all recommend a combination of light levels, and they tend to agree that most living rooms are inadequately lit. With this in mind, don't be afraid to add lights where they will prove useful. Most living rooms are large enough to accommodate a variety of experiments in lighting, from recessed downlights to wall washing or grazing.

Uplights of various kinds can be combined with strip lighting behind valances, coves, and wall brackets to provide indirect light that reflects off the ceiling. In any decor table or floor lamps can define specific areas by creating small pools of light interconnected with areas of relaxing shadow. This tonal variation in room lighting is restful to the eye and contributes to the overall atmosphere.

The use of plug-in fixtures offers several advantages and allows you to experiment with different options since these fixtures are easily repositioned. The main problem is what to do with the cord. Running a cord across the floor presents a hazard; hiding it under the carpet is equally dangerous. This generally restricts the use of plug-in fixtures to the area along the sides of the room, where the cord can be safely routed along the baseboard.

Use dimmers and three-way bulbs to control the light level in the living room. This will minimize the need to shift portable fixtures from one point to another when you want to vary the illumination. If possible, group the fixtures and individual switches on different circuits for additional control. If this is not possible, connect them in such a way that they can be operated in groups by different switches. Switches should be conveniently located, especially if the room has more than one door.

Variation in levels of light within the living room—while providing sufficient illumination for all activities—allows the eye to rest.

Light itself can become art. Here, projectors are focused above the fireplace to create a flowerlike image.

Different styles make good use of different lighting, with more dramatic illumination in the contemporary room and a greater feeling of warmth in the traditional room.

This tiny track with only three fixtures illuminates a large mirror and two couches. The room is meant for conversation around the fireplace, so task lighting is not provided.

While the recessed lights wash the large piece of art and the objects on the table, a small halogen floor lamp stands ready to provide task illumination.

Even a wood ceiling can use recessed lighting, as long as the work is carefully planned.

Positioning floor lamps at each end of a couch defines the space and provides suitable illumination for reading or other tasks.

Fixtures are focused on the most interesting objects in these rooms. Light bounces off the ceiling and floor to provide general illumination.

DINING ROOMS

Although a dining area is used most frequently for informal meals, it should be capable of assuming an elegant, festive air. The focus should be on the table and the lighting should show both the table and the food to best advantage. The lighting should also create a relaxed and convivial atmosphere.

Concentrate direct lighting over the table with downlights, pendants, or a chandelier. None of these light sources should affect the walls or ceiling. Long rectangular tables may need two or more pendants to illuminate them adequately. Position fixtures just above eye level so that they will not impede access to the table. Make sure that the fixture design will not cause glare.

Rise-and-fall fixtures can be adjusted to the precise height you want, and they are easy to clean. Avoid spotlights and diffusing fixtures for dining rooms. Spotlights are hard to position correctly with a group seated around the table. Diffusing fixtures tend to be a bit too bright.

When candlelight is desirable, it's better to use the real thing instead of lights that simulate candles. Make sure that the candle flames are properly shielded from drafts and positioned to prevent singed arms or sleeves, as in a simple candelabrum. To be most effective, candlelight should be combined with supplementary electric lighting. By itself candlelight is too dim to provide adequate illumination, but it creates a decorative touch when combined with faceted glassware and polished silver.

Separate fixtures over the buffet or along the wall will supply useful general lighting. This reduces overall contrast in the room without detracting from the table. Any accent lighting should be kept within correct proportions. Light concentrated on a wall near the table, for example, will be distracting and may be uncomfortable.

Dimmers can be used effectively in dining rooms. Bright light is useful for setting the table or seating guests: Once the food is served and the meal begins, lower the light level to create a casual, relaxed atmosphere.

A chandelier can easily become the focal point of the room, but it must fit the table and be properly positioned.

This large formal dining room is lit for dinner parties, displaying glassware and art, and focusing light on the table.

The soft lighting in this dining room provides sufficient light for eating and conversation, while allowing diners to see out onto the patio and into the night.

BEDROOMS

The word bedroom evokes restful slumber, which requires no light. However, the bedroom is more than simply a place to sleep. You dress in this room and store clothes in the closets. You may even use the bedroom as a quiet place to relax when other rooms are filled with activity. Because a bedroom plays many roles, the lighting should be carefully planned.

Bedrooms in older homes usually have a permanent fixture hanging from the center of the ceiling. Most bedrooms in modern homes have no light fixtures at all. Both approaches leave something to be desired. A centrally located hanging pendant can get in the way if it is hung too low. It limits the placement of the bed, and it may leave the corners of the room dark and gloomy. Bedrooms that have no permanent light fixtures do not give you the convenience of switching on the light when you enter the room, leaving you open to the risk of stumbling in the dark.

Bedrooms are good places to indulge yourself with imaginative and decorative fixtures, but don't go overboard. The bedroom can have a seductive atmosphere without depending on whimsical fixtures, colored lights, or other cliches often associated with fantasy.

Bedside lighting choices include small table lamps, adjustable wall-mounted tungsten lamps, and a valance on the wall above the bed. Besides blending into the decor, the bedside fixture should provide ample illumination for reading, and it should be easy to control from the bed. Couples should have separate fixtures, each of which can be used without disturbing the other party.

Good task lighting also should be provided at mirrors above a dresser or dressing table. Table lamps are a favorite choice, but choose the shade carefully for function as well as decor. It must provide enough light for general grooming or for applying makeup. Use a translucent shade at least 2 inches higher than the top of the bulb for best results. Narrow, opaque shades tend to illuminate the top of the dresser instead of the user's face. Full-length mirrors can be lighted by wall fixtures on each side. Be sure to light the closets well enough so that you can see the clothes.

Recessed lighting and a combination of wall and table lamps can be controlled separately to provide a variety of lighting options for this pleasant, calming bedroom.

A combination light and ceiling fan is well up out of the way, and small table lamps blend unobtrusively into this oh-so-feminine bedroom.

All lighting in this bedroom is soft and diffused, creating a very restful atmosphere.

Suspended 12-volt fixtures are the main source of light in this bedroom.

Recessed fixtures highlight paintings, table lamps permit reading in bed, and sconces provide a low level of illumination.

CHILDREN'S ROOMS

In lighting a child's room, think safety and growth. Use fixtures and outlets that will protect the curious child from getting a dangerous shock. For young children, childproof outlets and wall–mounted, sealed fixtures that give off little heat are best.

The uses to which a child's room is put change more quickly than is true of other rooms, so provide for these changes when you design the lighting plan. In a room for a baby or a very young child, lighting should be convenient for the parents to use. A dimmer adds flexibility, providing bright light for changing diapers and dim light for late-night feedings. Decorative night-lights give little children a sense of security.

With growth comes the need for greater flexibility. Make sure that the play area is sufficiently well lit to prevent eyestrain. Some form of reading lamp should be provided. Bunk beds need adequate light for both levels, and each level should have its own independent switch.

Older children will need good task lighting for homework and hobbies, as well as general lighting for those times when they just want to be alone. Clip-on task lamps work well for teenagers, since they can be moved around easily as needed. Diffuse light from wall brackets or strip lights over built-in bookcases will supplement the general light level and accent record or book collections.

What was once a child's room has changed with its growing occupant to become the elegant bedroom of a teenage girl.

STUDIES AND LIBRARIES

A study or a library may be a separate room or part of some other room that has been set aside for quiet concentration. In either case it requires general illumination combined with adjustable task lighting. There must be relatively shadowless illumination for typing, writing, bookkeeping, and so forth.

Easily adjustable task lights can be repositioned as required, while clip-on lamps will save desk space. There should be enough general illumination to prevent strong contrasts between the work area and its immediate surroundings. A floor lamp that can be adjusted to provide downward light, a ceiling-mounted downlight with a lens, or a pendant fixture that diffuses are all popular choices.

The reflectance of the room and its furnishings plays a part in determining light levels. Lighting above a dark wood desk is more effective when the desk top is covered with a white blotter. If the room contains many bookcases and books, try wall washers to increase the general light level while accenting the collection. The traditional atmosphere of a library can be maintained by using shaded table lamps or floor lamps.

Using a computer in a study presents lighting problems similar to those posed by a TV screen. Keep the contrast between the screen and its surroundings moderate, and avoid direct lighting that will cause reflected glare. Also consider the placement of the computer as it relates to windows. The window should be on your side as you sit at the computer, not in front or in back of you.

What was once a walk-in closet has been cleverly converted to a small but most useful home office.

Top: This library looks as if it could be in an English country manor, but it uses modern reflected general lighting bounced off the ceiling. The floor lamps are positioned for reading.
Bottom: This office keeps everything close at hand. It uses side lighting to avoid glare on the computer monitor.

GREAT ROOMS

Lighting areas designed for open–plan living require careful thought. If the room has an open beam or cathedral ceiling, recessed downlights with dimmers can provide good general lighting.

A more daring option is a suspended track system wired to allow general lighting on one circuit and accent lighting on another. Rise-and-fall ceiling pendants positioned at strategic points are less adventurous, but they contrast effectively with uplights or wall fittings.

After you have provided for the general lighting, block out the area into separate activity centers. Use individual fixtures to create interconnecting pools of light that can be related to one another or treated separately as desired. This will add to the functional use of the room and make it easy to change the mood or the atmosphere.

Portable fixtures find favor in great rooms, since they are highly flexible and can be positioned as the occasion requires. Don't overlook the value of adjustable task lighting, which can serve a dual purpose when necessary by providing indirect background light bounced off one of the walls.

The open spaces of great rooms allow for considerable freedom in the arrangement of both furniture and lighting. Individual fixtures highlight each seating area and art object.

HALLWAYS

Hallways must be lighted to allow the eye to adjust as you move between rooms. The lighting must also be bright enough to illuminate stairs, corners, or other potential hazards. Hallways can also present excellent spaces for displaying artwork, which should be highlighted for best presentation.

Recessed ceiling lights fitted with either tungsten bulbs or miniature fluorescents and diffusers are a good choice. They can be used in short or long hallways. Space the lights carefully in long hallways to avoid creating an alternating pattern of light and shadow. Another possibility is to use wall sconces that direct the illumination toward the ceiling.

When hallways are used to display pictures or artwork, track-mounted spotlights are a convenient solution.

A switch at each end of the hall offers both convenience and safety, allowing you to turn the lights on and off as you pass from room to room. If there are closets in the hallway, make sure that they have ceiling fixtures of their own.

Even though it is only a transitional space, a hallway can be lighted with a fine sense of style, as this example demonstrates.

Light seems to radiate from the unique vaulted ceiling in this hallway, with lighter colors and brighter light beyond.

Whether it consists of 2 or 20 stairs, a stairway must be sufficiently well lit to permit safe movement up or down, even by those with failing eyesight.

Avoid flat general lighting and use strongly directional light to create contrast between the horizontal treads and the vertical risers. Illuminate the edges of the steps with a downlight installed near the top of the stairway. A softer light at the bottom of the stairs will help to define the depth of each step, but it should be carefully shaded to prevent glare when seen from above. The additional light provided by a wall bracket can be useful in the case of long stairways.

If you are remodeling an area containing a stairway, consider lighting the stairs more subtly. Hide fluorescents under or behind the handrails; install low-wattage recessed lighting in the wall above the risers; or conceal ministrip lighting under the treads. Each of these alternatives will provide adequate light and add a touch of the unusual as well.

Unusual fixtures help to make this stairway not just safe but interesting as well.

Although the lighting is subdued, each step is visible to ensure safe passage up and down the stairs.

SELECTING LIGHT SOURCES

Interior light sources are generally classified according to the way in which they produce light. Incandescent light results from passing an electrical current through a tungsten filament in a bulb. Fluorescent, neon, and high-intensity discharge (HID) light is produced by electrically charging the gas or phosphor coating inside a tube.

To decide which light source is most appropriate for a given application, you must consider the cost, energy requirements, color appearance and rendition, bulb shape, and light distribution.

The exact order in which you rank these factors will depend on your own personal priorities, but each factor should play a part in the final decision.

A view through several rooms reveals correct placement of a hall pendant up and out of the way and a chandelier carefully centered over the dining table.

WHICH LIGHTS ARE RIGHT FOR YOU

Some types of light sources are perfect for accent lighting but wouldn't work at all for general lighting. Others may be best for task lighting. Some may seem expensive at first, but save electricity year after year. Others may have a low initial cost but eat up power at an alarming rate.

Cost

Whether you are engaged in new construction or in remodeling, you should break down the cost of any lighting system into two parts. The first is the cost of the initial installation; the second is that of operation and maintenance. For example, fluorescent tubes and tungsten-halogen bulbs are more expensive to buy than tungsten bulbs but have a far longer life. They also differ considerably in operating cost. Tungsten-halogen bulbs use about the same amount of electricity as tungsten bulbs. Fluorescent tubes use much less electricity and thus cost less to operate.

Remember that after the lighting system has been installed, approximately 80 to 90 percent of the expense will be for the electricity used to run it. Only 10 to 20 percent will be for replacing bulbs or fixtures.

Energy Requirements

All light bulbs are rated in watts. This rating is a measure of the amount of electrical power they consume. A 100-watt bulb produces a brighter light, but also consumes more energy and thus costs more to operate, than a 25-watt bulb. However, it is generally more efficient to use one incandescent bulb with a high wattage rating than it is to use several low-wattage bulbs. For example, two 60-watt bulbs will produce the same amount of light as a single 100-watt bulb, but will require 20 percent more power to do so. In general, the fewer lamps you use to obtain a given quantity of light, the more economical your lighting system will be. Low-voltage bulbs produce brighter light for fewer watts when compared with incandescent bulbs.

Energy efficient or long-life incandescent bulbs do consume less energy and last longer than ordinary incandescent bulbs, but they give off somewhat less light during their life. In addition, an energy efficient bulb consumes only about 5 watts less than a comparable standard bulb. Depending on the cost of electricity in your area, the money you save by using these bulbs may only offset the higher price you pay for them. If you are conservation minded, however, the energy savings may appeal to you anyway.

Fluorescent tubes are the real power misers of the lighting world. A 40-watt fluorescent tube uses only about 40 percent of the energy required by a 100-watt incandescent bulb, yet it produces almost twice the light. The fluorescent tube also gives off far less heat, and it lasts up to 10 times as long as a comparable incandescent bulb.

The life of an incandescent bulb is considerably shortened by incorrect line voltage. If the electric company is not regulating the voltage properly and the power coming into your home is actually higher than 120 volts, the bulbs will burn brighter but will not last as long. For example, a 100-watt tungsten bulb may be rated by the manufacturer for 1000 hours at 120 volts. However, if the line voltage is 130, the bulb will last only two thirds of its rated life, or approximately 667 hours.

Color Appearance and Rendition

The quality of light produced by a given source is related to its color temperature. Color temperature determines both how you see the light (warm or cool, yellowish or white) and how the light affects the object it illuminates. The color appearance of any light source can be altered by passing the light through lampshades, screens, window coverings, or other filtering materials.

Incandescent light falls on the lower end of the color spectrum, giving it a pleasant cast, warmer than that of sunlight or fluorescent tubes. It is considered flattering to skin tones. Although incandescent light accentuates yellows and oranges, it makes blues appear grayish and flat.

The light from standard fluorescent tubes is higher on the color spectrum, and contains more blue. Its effect is almost exactly opposite that of incandescent light. Fluorescent light is considered uncomplimentary to skin tones; it tends to accentuate blues and to make yellows and oranges seem grayer. However, the new

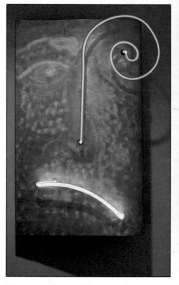

Neon can be used in a variety of ways. Here it is worked into a wall sculpture.

color-corrected fluorescents produce a warm light similar to that of incandescent bulbs and have overcome many of the shortcomings of the earlier fluorescent bulbs.

Bulb Shape

Tungsten bulbs are manufactured in a wide variety of shapes. Besides the traditional pear-shaped household lamp, they are available as globes, tubes, teardrops, and flames. The decorative forms are generally used in fixtures in which the bulb is visible and may be designed to show off the filament.

Halogen bulbs all have unusual shapes, ranging from ovals the size of a peanut to small tubes about half the length of a pencil. They are generally designed for horizontal use, to prevent the filament from fracturing. The most common unit is a tube with the filament connected to an electrical contact at each end. Of course, these unusual shapes cannot be inserted into a standard light socket, but halogen bulbs are also manufactured with an outer glass jacket and a standard screw base. This bulb-within-a-bulb design can be used in an ordinary socket.

Fluorescent tubes are usually tubular or circular, but some of the newer ones are shaped like globes or cylinders. These new designs have a U-shaped fluorescent tube within the globe or cylinder and can be inserted into a standard light socket. The starter and ballast required to operate fluorescent tubes generally make it impossible to use them in small fixtures or restricted spaces. However, adapters containing the components allow compact fluorescents to be used in many home fixtures.

Types of Bulbs

The bulbs in your lighting systems are an important part of your lighting plan. Using a type of bulb not meant for a fixture can be inefficient and sometimes dangerous.

Tungsten Bulbs

A tungsten bulb produces incandescent light. It contains a coiled tungsten filament inside a container made of soft glass. During manufacture, the air inside the glass is evacuated through an exhaust tube and replaced with various inert gases such as nitrogen and argon. These gases improve the quality of the light produced and retard evaporation of the filament.

The filament is supported by a stem press, button rod, and support wires. When heated by an electrical current, the filament starts to glow, giving off a reddish light in addition to heat when its temperature reaches about 943° F (500° C). As the filament gets hotter, the light given off becomes more intense and changes in color. It becomes orange, then yellow, and finally almost (but never quite) white. This process takes place so fast that the human eye cannot follow it.

Because of the intense heat, the filament slowly burns up or evaporates. As this happens the particles of the oxidized filament build up on the interior of the bulb, causing it to turn black. The gases inside the bulb slow down this process, but the filament eventually becomes so thin that it breaks. When this happens, we say that the bulb has burned out.

Tungsten bulbs are ideal for general lighting, especially in bedrooms, living rooms, and dining areas, where a feeling of warmth is part of the ambience. The wide variety of fixtures available (see the fifth chapter) make for even greater versatility.

Tungsten bulbs can also utilize a variety of controls, from straight on-off switches to dimmers. The latter are silent and relatively inexpensive, and they lend themselves well to many uses.

However, tungsten bulbs have a relatively short life (500 to 1000 hours). They are less efficient than other types of bulbs and they use considerably more electricity. Only about 6 percent of the electricity consumed by tungsten

Crown-Silvered Bulb With Parabolic Reflector

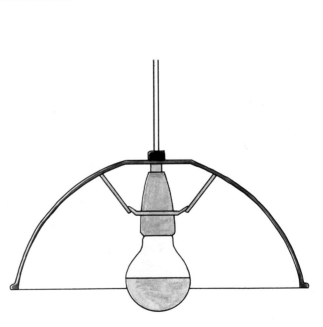

Used with a parabolic reflector, the crown-silvered bulb creates a decorative effect and produces a controlled beam of indirect light with a minimum of glare.

bulbs is used to produce light; the rest generates heat, which must then be dissipated safely. For this reason the choice and placement of the fixture often limit the size of bulb that can be used.

A-Bulb

The most widely used household lamp is the pear-shaped common light bulb. It contains a single filament. Clear A-bulbs produce the most light but tend to create glare. For this reason they are used mainly in fixtures with diffusers. Frosted finishes give a softer light. The inside of the bulb is etched with acid to produce the frosting. Soft white bulbs have a fine coating of white silica, which obscures the filament and diffuses the light. They create less glare than frosted bulbs, and their light softens shadows better.

Some A-bulbs have a soft pastel or deep pigment coating, which produces a colored illumination. Others have a white matte outer coating, which diffuses the light more efficiently for close work.

Three-Way

The three-way bulb resembles the A-bulb but contains two filaments. The filaments can be used separately or in combination to produce a choice of three lighting levels, all controlled from the same point. Three-way bulbs are most valuable in multiuse areas. One level can be used for a subdued background light, another level for a medium light, and both levels combined to provide a bright light. This bulb requires a special socket and switch.

Long-Life

The long-life bulb contains a longer filament than the A-bulb or the three-way. This is a compromise design; it uses slightly less electricity but also produces somewhat less light than an A-bulb. The higher

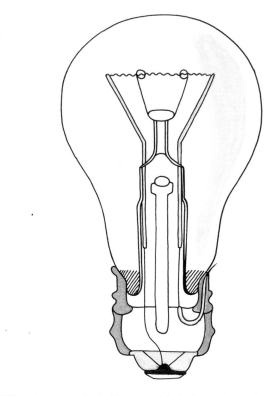

Cutaway of Tungsten A-Bulb

This cutaway of a typical tungsten A-bulb shows its construction, which has changed little since Edison's day.

The Incandescent Lamp

The incandescent lamp was invented in 1878 by Joseph Swan, an Englishman who had developed a vacuum glass envelope containing a carbon filament. Swan's bulb, however, had a very brief lifespan, and the first practical light bulb did not appear until 1879. In that year Thomas Edison produced a bulb containing a filament of carbonized silk thread, which lasted for 45 hours before burning out.

Edison and Swan had worked independently to achieve essentially the same results, differing mainly in the way in which their bulbs were connected to an electrical supply circuit. Edison's bulb used a threaded brass cap that screwed into a corresponding socket. Swan had devised a bayonet base with lugs, which was depressed in its socket and then turned to lock it into place. Interestingly enough, these remain the two most

popular types of connector to this day. By 1880 the two men had joined forces and formed the Edison and Swan United Lamp Company.

The process of refining the incandescent lamp continued, with an emphasis on more durable filaments. Carbonized bamboo and various metals were tried, but real success awaited the introduction of tungsten. Pressed tungsten was first used in 1907 and replaced by drawn tungsten in 1910.

At the same time experiments led to replacing the vacuum in the bulb with various inert gases that would further extend filament life. In 1913 the appearance of a double-coiled filament completed the development of the incandescent lamp. Although the size, shape, and appearance of the glass envelope continued to evolve, the basic light bulb remained essentially unchanged until tungsten-halogen lamps appeared in 1964.

50

initial price offsets most of the reduction in operating cost, so you won't save much by using these bulbs. However, their long life makes them particularly well suited for use in hard-to-reach fixtures and locations.

Using 130-volt bulbs is an alternative to using long-life bulbs. Just as excessive line voltage shortens the life of a bulb, using a bulb with a rated voltage slightly higher than the line voltage will lengthen its life. Using a bulb rated at 130 volts will more than double the life, but the light output will be about 75 percent that of a 120-volt bulb. These bulbs cost a little more initially but operate at a slight savings.

Reflector

In addition to the three major types of tungsten bulb, there is a variety of specialized types. The funnel-shaped reflector (R) bulb is the mainstay of interior spotlighting and floodlighting. This bulb is made of blown glass. The interior is coated with a white or silvered material, leaving only the crown clear. This coating acts as a mirror to reflect a controlled beam of highly intense light through the crown. R-bulbs are most useful in track lights or recessed fixtures. They must be used in heat-resistant ceramic sockets, since they create far more heat than do ordinary tungsten bulbs.

The elliptical reflector (ER) is similar in design to the R-bulb, but it delivers a concentrated beam with a coverage between that of a spotlight and a floodlight. The shape of an ER-bulb and the type of coating used combine to focus the light beam about 2 inches in front of the bulb before it expands. A 75-watt ER-bulb will deliver the same amount of light as a 150-watt R-bulb, but in a somewhat smaller area. Heat-resistant ceramic sockets are also required for ER-bulbs.

The parabolic aluminized reflector (PAR) is a pressed-glass bulb with a hardened prismatic glass front. It is available with either a wide beam for floodlighting or a concentrated beam for spotlighting. More durable than R- or ER-bulbs, a PAR can also be used outdoors without additional protection. Cool-beam versions of the PAR-bulb contain a dichroic reflector, which reflects the light forward while transmitting the heat backward. Although this removes up to 75 percent of the radiant heat from the light, all PAR-bulbs require heat-resistant sockets and fixtures.

Crown-Silvered

Another A-bulb design has a silvered crown that masks the filament. This reduces glare and produces indirect light. For maximum effect the bulb must be used in an appropriate parabola-shaped reflector. The internally silvered crown of the bulb directs the light back onto the reflector surface, which throws a controlled beam into the room.

Chandeliers often use small flame-shaped bulbs to simulate candles.

Low-Voltage Reflector Spot

Relatively new to interior lighting, the smaller version of the standard R-bulb is designed to operate at 12 or 24 volts instead of 120. A transformer must be used to step down the household voltage. You can use a separate adapter instead, but the transformer is generally part of the fixture. With track lighting a single transformer can be used for several units. The transformer raises the cost of a low-voltage fixture, but if the lighting plan is well thought out, low-voltage lighting can be economical in the long run.

Because it lends itself to precise direction, the low-voltage reflector spot is ideal for accenting a small area. Since the bulb runs on lower voltage, it can use a smaller filament to create a more concentrated light beam. It also eliminates the problem of scorch marks that can result when a normal-voltage spot is installed too close to a wall or ceiling.

Ministrip Lights

Small bulbs enclosed in a flexible plastic strip, ministrips have many uses. They provide concealed lighting, picture lighting, staircase illumination, and decorative lighting around a mirror or the headboard of a bed. Ministrip lights are available for standard or low-voltage applications.

Purely Decorative Bulbs

Although many tungsten bulbs can be used for decorative purposes, a large assortment of specialty bulbs is also available. These include Christmas tree lights and frosted minibulbs covered with colored glass or plastic fragments. They are

Low-Voltage Halogen Bulb and Fluorescent Bulbs

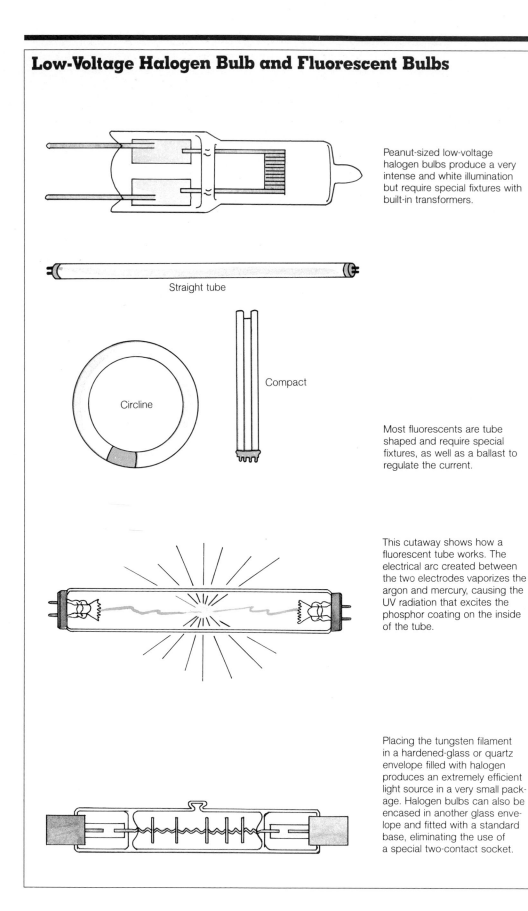

Peanut-sized low-voltage halogen bulbs produce a very intense and white illumination but require special fixtures with built-in transformers.

Straight tube

Circline

Compact

Most fluorescents are tube shaped and require special fixtures, as well as a ballast to regulate the current.

This cutaway shows how a fluorescent tube works. The electrical arc created between the two electrodes vaporizes the argon and mercury, causing the UV radiation that excites the phosphor coating on the inside of the tube.

Placing the tungsten filament in a hardened-glass or quartz envelope filled with halogen produces an extremely efficient light source in a very small package. Halogen bulbs can also be encased in another glass envelope and fitted with a standard base, eliminating the use of a special two-contact socket.

generally used for special occasions, such as holidays or parties.

Tungsten-Halogen Bulbs

Originally designed for use in floodlighting and displays, tungsten-halogen bulbs are becoming popular in interior applications. Tungsten-halogen bulbs maintain their light output better than other incandescents, losing only about 5 percent of their rated lumens during their life, as compared to about 20 percent for other types.

Although a tungsten-halogen does not look like an ordinary tungsten bulb, it works on a similar principle. A heat-resistant quartz envelope contains the tungsten filament and pressurized halogen gas. When the filament is heated by electricity, it gives off light. The high operating temperature combined with the halogen gas surrounding the filament produces a brilliant white light.

In an ordinary tungsten bulb, as the filament burns its vaporized particles are deposited on the inside of the glass. This darkens the bulb and reduces its light output. A regenerative cycle prevents this from happening with a tungsten-halogen bulb. The filament particles combine with the halogen to form a compound that redeposits the tungsten on the filament. This process continually renews the filament as the halogen is released to repeat the cycle. For this reason a tungsten-halogen bulb can last as long as 4000 hours.

Using a dimmer control to run the bulb at less than full voltage will temporarily interrupt the regenerative cycle. This causes the bulb to darken gradually, just like an ordinary tungsten bulb. When this happens the cycle can be restored by applying full voltage until the surface of the bulb is clean.

Tungsten-halogen bulbs must be handled carefully. Skin oil from fingerprints will weaken the surface, causing the bulb to fail prematurely. Since the bulb can heat to more than 500° F, a premature failure may cause it to explode. You should wear gloves or use a clean cloth when changing a halogen bulb, unless the manufacturer has provided a protective glass sleeve or jacket.

Since the halogen bulb gives off a higher level of heat than an ordinary tungsten bulb, its fixture must be properly designed to dissipate the heat satisfactorily. The fixture must also be positioned carefully to avoid heat damage to nearby objects. When buying a halogen fixture, make sure that it is UL approved. Shading is also important because of the intensely bright light and the possibility that the bulb may explode.

Their brilliant light and small size combined with the compactness of halogen fixtures make tungsten-halogen bulbs ideal for indirect or background lighting. They are also useful for task lighting, and they can be focused to provide a narrow beam for spotlighting.

Low-Voltage

Two types of low-voltage halogen bulb are popular. Peanut-sized bulbs are used in desk lamps with built-in transformers. Reflector (R and PAR) bulbs and minireflector (MR-16) bulbs are used as spotlights or in recessed downlights and low-voltage track lights. Some of them contain a dichroic reflector that projects the light forward and the heat backward to create a cool beam. These must be used in open fixtures or recessed fixtures specifically designed for them, because of the heat they throw back toward the socket.

Fluorescent Tubes

Unlike incandescent light, which is created by heating a tungsten filament with electricity, fluorescent light is produced by using electricity to stimulate a phosphor coating on the inside of the tube. The tube contains tiny droplets of mercury and an inert gas such as argon or krypton. The caps on each end, used to support the tube in its holder, contain an electrode or cathode. Current entering the tube through

Put the Pinch on Energy Costs

Since the energy crisis of the 1970s, manufacturers have developed a wide range of energy-saving bulbs as direct replacements for the standard incandescents. In many cases the newer bulbs cost the same as the ones you've been using, and the difference in performance is slight compared to the savings in energy. Here are some other easy ways to cut your lighting costs.
• Check incandescent bulbs and replace any that are darkened. Don't wait until they burn out. Although such bulbs are still "burning," they use the same amount of power as a new bulb but give off less light.
• Accumulated dust and dirt absorbs light. Clean bulbs, globes, and fixtures on a regular basis, especially if there is a smoker in the house.
• Use low-wattage reflector bulbs instead of A-bulbs in directional fixtures. A-bulbs trap light in such fixtures; reflector bulbs direct the light where you want it.
• Use reading lamps with three-way sockets and keep them on *low* when they are turned on but are not being used for a specific purpose.
• Turn lights off when they are not being used.
• Don't use higher wattage bulbs than necessary. If you have a 75-watt bulb in a hallway, try a 60-watt bulb instead.
• Use fewer bulbs if possible. A single 75-watt bulb provides 1210 lumens; three 25-watt bulbs give off a total of 705 lumens. In this case you'll get almost twice the light output for the same amount of energy, and the single bulb costs less than the three lower wattage bulbs.
• Check the information on the package when you buy bulbs. Off-brand bulbs may seem inexpensive, but if they're not giving you a lumen output equivalent to brand name bulbs, you're not really saving anything.

the caps creates an electrical arc between the two electrodes. The heat of this arc vaporizes the mercury, causing it to give off ultraviolet (UV) radiation. This UV radiation reacts with the phosphors in the coating, which glows or fluoresces, producing a bright light that radiates from the entire tube surface in all directions. The high level of diffuse light produced by a fluorescent tube results in flat illumination with no shadows, making it ideal for area lighting. The color of the light given off varies depending on the composition of the phosphor.

Standard 120-volt electricity arrives at the fluorescent fixture, is decreased as it passes through a ballast or transformer, and enters the tube. The ballast also regulates the current to reduce electrode erosion.

There are three types of fluorescent: preheat, rapid-start, and instant-start. In the preheat tube, the starter and ballast are separate; this causes a brief delay before the tube reaches full light output. The starter is built into the ballast on rapid-start models. Instant-start models use no starter and can be distinguished by their single-pin

tube. Instant-start tubes reach full light output immediately.

The starter in preheat tubes must be replaced periodically for satisfactory operation.

Rapid- and instant-start tubes consume a little more electricity than preheat models but give slightly more light. All fluorescent tubes eventually fail from electrode erosion rather than exhaustion of the mercury vapor or phosphor coating.

Fluorescents are very efficient, use only one fifth to one third the electricity of a comparably bright incandescent bulb, and last 10 to 20 times as long. Energy-saving rapid-start tubes are rated as high as 20,000 hours. Colored tubes

lend themselves well to accent or mood lighting.

The noise and flicker often associated with fluorescent lighting is caused by lack of maintenance. An excessive humming noise indicates that the ballast should be replaced; a flicker indicates a bulb that has deteriorated beyond its rated life and needs to be replaced.

Although they are highly efficient and long-lasting, fluorescents have definite limitations. The harsh, cool, and unflattering light of earlier tubes restricted the use of fluorescent light primarily to the kitchen and the bathroom. In recent years, however, triphosphor technology has produced a variety of color-

corrected tubes that come close to matching incandescent light. Yet even these modern tubes cannot be focused, are rarely used with dimmers due to the expense, and provide a flat, shadowless light.

Tubes

The traditional design used to produce fluorescent light is the tube. Tubes come in lengths from 5 to 96 inches and in three diameters, but they require special fixtures and a ballast for operation. The new color-corrected tubes approximate incandescent light. Tubes are useful for both area and indirect illumination. They bathe surfaces with light, reducing apparent texture,

Read the Box

The basic information you need to buy light bulbs of any kind is provided right on the box or carton in which the bulb is packaged.

Watts
This is the number that most people look for. It tells how much power the bulb consumes but nothing about the light output.

Average Lumens
This is the amount of light given off by the bulb when it is new. For reference, the candle on a dining table gives off about 12 lumens. Remember that a tungsten bulb will dim by up to 20 percent as it gets older.

Average Life
This tells how many hours the manufacturer expects the bulb to last. The number given is generally the median—that is, it is the number of hours by which half of the bulbs of this type will have burned out.

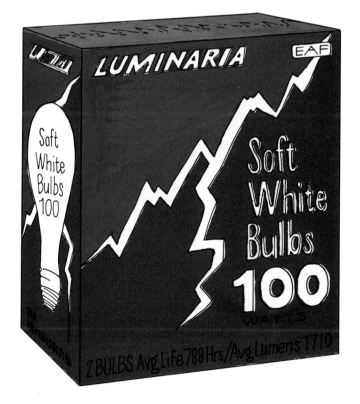

shadows, and outlines. Specially coated pastel tubes have decorative uses.

Compact Tubes

The double or U-shaped compacts are 5¼ to 7 inches long—small enough to replace incandescent bulbs in some applications. The ballast may be housed in the screw base or in an adapter between the incandescent socket and the tube. New fixtures are being designed specifically for these lamps, incorporating the ballast into the design. Compact tubes are suitable for ceiling and wall fixtures, as well as pendants and table lamps.

Miniatures

Available in a variety of globular or cylindrical shapes, miniature fluorescents are tubes inside a bulb. They fit standard lamp sockets and come in several color temperatures. Since the ballast is built into the base, a miniature fluorescent can replace an A-bulb directly without installing an adapter or modifying the circuit. They are highly economical for general lighting.

Circles

Circular tubes are designed for use in special ceiling fixtures, but they can be used with an adapter in other applications. They are available in 6-inch to 12-inch diameters. Their electrode pins are located on a rotatable collar for easy installation.

Neon

This type of lighting consists of a length of clear, coated, or tinted glass tubing with electrodes at each end. After the air is removed from the tubing, it is filled with a rare gas such as neon or argon and sealed. Sending high voltage through the electrodes excites the gas, causing it to glow.

The color of the light given off is determined by the gas and tubing used; over 40 colors are possible. The basic color of pure neon is a warm reddish orange; pure argon produces a bluish light. All other colors are produced either by putting additives in the gas or by tinting the tubing.

Although neon lighting consumes very little electricity and produces almost no heat, it does require a transformer to step up the voltage to the 2000 to 15,000 volts necessary for operation. The ¼-inch to ⅝-inch diameter of the tubing allows it to be bent into intricate designs for decorative or accent lighting. As long as the tubing remains sealed, neon lights will last for decades.

Neon is used in some flame-shaped bulbs with candelabrum bases. These bulbs flicker when lit, simulating candles. However, the most common use of neon lighting in the home is in the form of neon sculptures, such as a cactus or flower. Recently neon appears to be making a comeback, with designers incorporating it into telephones, clocks, and paintings and using it to accent niches and other recessed areas.

High-Intensity Discharge (HID) Bulbs

The last category includes a variety of high-intensity discharge (HID) bulbs. These are filled with sodium, metal halides, or mercury vapor. Light is produced by electrically exciting the pressurized gas in the bulb. HID bulbs require special fixtures and ballasts, and they do not produce light immediately when they are turned on. HID bulbs are used mainly for outdoor security or landscape lighting.

Cabinets and display cases should be lit to set off the contents to best effect. Here, silver and glassware gleam.

SELECTING LIGHT FIXTURES

To select the best light fixture for your purposes, you must first determine what kind of light you want, how much light you need, and where and how the fixture is going to be positioned and used. Of course, the fixture should also harmonize with your furnishings. The fixtures you choose are as important as the light source in determining the character of your lighting. The previous chapters gave you the information you will need to choose the kind and quantity of light you want. This chapter will help you to select your fixtures.

Kitchen lighting can be both attractive and utilitarian. These fixtures provide efficient task lighting for food preparation.

FIXTURE FUNDAMENTALS

Years ago it was fairly easy to choose a light fixture. There simply wasn't that big a selection. Today, however, the variety of light fixtures is mind-boggling. Selecting a single fixture can take considerable thought and effort. Locating and choosing a whole houseful of fixtures can easily become a full-time occupation.

Selecting Fixtures

In large cities lighting labs enable you to compare the effects of different fixtures and different light sources. Take full advantage of any lighting labs or showrooms in your area. The greater your needs, the more help such places can be.

Function

Any lighting system should include a combination of overall illumination and strong directional light. Visualize the effect of the entire system before you choose any fixtures, and decide how each light in the system will be related to the others. Make sure that any fixtures you are thinking of buying will put the light where you need and want it.

Nondirectional fixtures must flood the area with a suitable intensity of illumination. Directional fixtures must take bulbs with sufficient wattage to send the light where you intend it to go. When the two are used together, the resulting light should be well balanced, and you should be able to vary its degree and intensity to suit your needs.

Aesthetic Appeal

Taste, of course, is a personal matter. Generally the fixtures should harmonize with the overall decor. When you install permanent fixtures in a house, consider how they will affect the resale value. Radical designs may appeal to you but drive away potential buyers. If you are renting be sure not to do anything that violates your lease. The selection of fixtures is so great that you should be able to please yourself and augment the value of your home, or satisfy your landlord, too.

Quality

The quality of any fixture depends on materials and construction. For example, consider the ever-popular brass fixture. You probably think of it as being made of solid, heavy castings. However, it can also be stamped brass, spun brass, or simply brass plate. All of these brass fixtures will look

alike, yet each may require different care to retain its appearance. Make sure you know what you are buying and how to care for it.

The way in which a fixture is assembled is important both to its useful life and to its safety. Some of the components can be seen; others are hidden. For this reason it may be difficult to judge the quality of construction. The best assurance you can have regarding safety is the UL label, indicating that the fixture has been approved by the Underwriters' Laboratories, Inc.

Compatibility

Any fixture should be carefully matched with its light source to avoid creating a potential hazard. Manufacturers specify the recommended maximum wattage of bulb that can be safely used in each fixture. If you exceed this recommendation, you risk overheating the connections or the socket and creating a fire hazard. If you require more light than the fixture can deliver safely, choose a different fixture.

You should also match the bulb type to the fixture design. For example, a parabolic reflector takes a crown-silvered bulb, not a standard A-bulb. Using a bulb not designed for the fixture wastes its light output and may create a safety hazard. Using a cool-beam PAR-bulb in a standard spotlight without a ceramic socket invites overheating, for example.

Size

Don't be misled when you look at the fixtures on display in a store. Seen in the context of the lighting display, a fixture may appear to be just the right size. When you get it home among your own furnishings, however, it may seem to have suddenly grown.

This problem of scale can be easily avoided by measuring the fixture in the store. Once at home, check for proportion by using a box or carton of a similar size to approximate the fixture in the planned location. If it does not seem right and the fixture is a standard design, the store may be able to supply it in other sizes.

Flexibility

If you change your mind or your decor, will the fixture you bought this year be usable next year? Here movable fixtures, such as floor or table lamps, have a definite advantage. If you can't use them in one room after you've redecorated, they may well find a useful place in another—a virtue that ceiling and wall fixtures obviously lack. You can modify these more permanent fixtures to a limited extent by changing the globes or the trim.

Flexibility also helps to account for the popularity of track lighting. You can change the location of individual lamps in the track, and you can aim the light beams in a variety of directions.

Cost

There are two aspects to the cost of any fixture: initial price and operating expense. The first is easy to ascertain; the second may be a bit more difficult, but this information should be available from the dealer or manufacturer. A less expensive fixture can often do the job just as well as a costly one, and it may be of equal quality.

If you use the lighting system for extended periods, you may want to consider investing in a low-voltage or other energy efficient unit. Although the initial price is higher, it may prove more cost-effective than the less expensive unit in the long run.

Maintenance

Light fixtures require periodic cleaning, if they are to look attractive and work properly. Fixtures used in bathrooms, kitchens, and work areas such as laundry rooms need the most attention and should be easy to clean.

Crystal, stained glass, acrylics, and other plastics used in decorative fixtures also require frequent dusting and cleaning. Complicated fixture designs can make this a time-consuming task. Do not install such fixtures in hard-to-reach places. Fabric lampshades tend to collect dust quickly. Textured fabric shades are especially difficult to clean.

One final word when it comes to maintenance: The design and location of the fixture can make changing the bulbs a major task.

Control of Light Distribution

The design of the fixture sometimes determines how the light is distributed. The simplest fixtures consist of a decorative bulb. These have little or no effect on light distribution. More complex designs modify and shape the light in a variety of ways.

Reflection

R-bulbs contain built-in silvered reflectors to control or redirect the light they produce. However, reflection is generally a function of the fixture itself. Placing a shiny (generally metallic) surface behind a bulb will bounce the light it produces into the room.

The shape and surface of any reflector determine its

Lighting for Effects

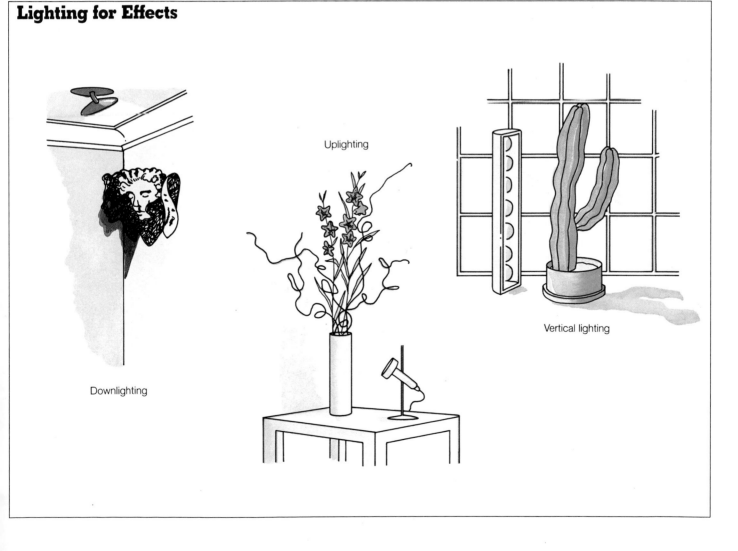

Downlighting

Uplighting

Vertical lighting

effect and its efficiency. A hemispherical reflector radiates the light outward in all directions; a parabolic reflector focuses the light into parallel beams. Dimpled or matte reflectors are less efficient than highly polished ones but have the advantage of diffusing the light enough to minimize glare.

Reflectors can be tinted to impart different hues to the light. They may also be detachable or directional to control the placement of the beam.

Diffusion

The most common method of regulating light distribution is to diffuse it with a white coating inside a tungsten bulb. When diffusion is used as a feature of fixture design, it generally takes the form of a translucent panel over the bulb.

Diffusion is determined by the placement of the bulb behind the panel. When the two are properly positioned, the light is spread over the entire surface of the diffuser, which hides the bulb. The result is a nondirectional and relatively shadowless light useful for general illumination.

Once commonly made of etched or opal glass, diffusers are now generally made of milky white plastic. Besides being inexpensive and less breakable, plastic diffusers are lightweight. This places less stress on the fixture, and it

allows designers to indulge their creative fantasies in complex applications.

Refraction

The bending of light as it passes through a glass lens or prism is called refraction. Refraction can be used to spread and soften the light beam or to deflect it in a particular direction. A common example is the crystal chandelier.

Refraction is often used in combination with reflection or diffusion. In a framing projector, for example, the bulb and reflector can each be adjusted to alter the angle of the light beam, thus changing both refraction and reflection. Molding tiny plastic prisms into a diffuser intensifies the level of brightness while keeping it nondirectional.

Baffles and Louvers

Reflectors, diffusers, and refractors modify the output of a light source directly by intensifying, spreading, or directing it. Baffles and louvers reduce or redirect the light output. They are particularly useful in minimizing unwanted light spillage or reducing glare.

Baffles generally take the form of grooved indentations on the inner surfaces of spotlights,

downlights, and other recessed fixtures. To work effectively, the grooves are finished in a flat or matte black that absorbs light. Baffles are used mainly to define the edges of the light beam, but they have an additional benefit. If the fixture is in the line of sight, baffles reduce excessive brightness on the sides, preventing glare.

Louvers are opaque or translucent metal or plastic slats arranged to reduce glare and to conceal the bulb. Louvers used in downlights take the form of concentric blades. In horizontal fixtures the blades are usually placed at a right angle to the axis of the bulb.

Shades

Many kinds of shades are used to modify and direct light. They may be part of the fixture, as in a spotlight, or separate, as with a table lamp. Most people think of a shade mainly in terms of its decorative function, but how it deals with light is really more important. This is especially true when you consider that almost all lights use some form of shade.

The size, shape, material, and positioning of the shade all affect light distribution. The size of the opening is one important factor. A cylindrical shade produces a narrower beam than a flared one. Shades

The task lighting above this range helps to make cooking safer and shows off the handcrafted tiles.

used on table lamps have two openings, one at the top and one at the bottom. This controls the distribution of light in two directions. The upper opening sends light toward the ceiling, where it is reflected back into the room as indirect illumination. Light passing through the bottom opening reaches a table or the floor directly. The table or floor, however, may reflect some of this direct light back into the room as indirect light.

A translucent shade allows diffused light to escape into the room; an opaque shade sheds light only from its top and bottom.

The material and color of the shade have a decorative value, but they also affect the quality of the light. To ensure that you are getting the effect you want, try before you buy. Put the shade on a lamp, turn the lamp on, and note the results. Also check the inside of the shade. The lighter the color, the more light it will direct into the room.

How the shade is positioned is important. With table lamps you can control the beam spread somewhat either by changing the harp or by using an extension or shade riser, available from lighting supply stores. Raising the bulb inside the shade narrows the beam. The exact amount of control possible with a table lamp depends on the shape of the shade and on the bulb.

Shade positioning offers similar control with ceiling-mounted pendants, but here you must be careful since the bulb can generally be seen from many more places within the room. If the shade is positioned correctly but you still have glare, it may be necessary to use louvers or some other type of diffuser.

Types of Shades

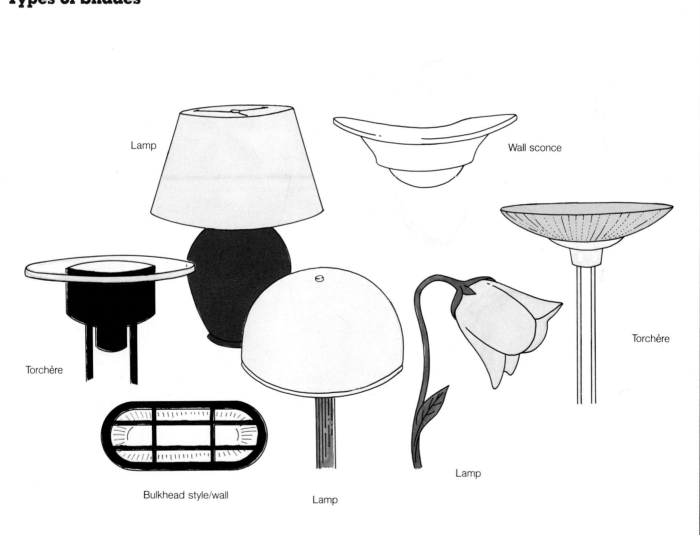

Lamp

Wall sconce

Torchère

Torchère

Lamp

Torchère

Bulkhead style/wall

Lamp

TYPES OF FIXTURES

There are fixtures to suit just about any application, budget, and space from tiny 12-volt systems to large architectural columns. Used well, fixtures can greatly enhance the appeal of your home.

Portable Light Fixtures

The many types of portables have several virtues, making them a popular choice in designing room lighting. They are inexpensive to buy, can be repositioned according to your needs or tastes, and can easily be taken with you if you move.

Table Lamps

Available in virtually any decorating style, lamps for the table generally consist of a decorative base fitted with a light socket. The shade may be included or sold separately. Many table lamps can be adapted to a new decor simply by changing their shades.

Table lamps make a decorative statement while delivering light where you need it. Since popular designs often come in various sizes, you can select one that will be correctly proportioned to your room.

The design of the shade is critical to the proper performance of a table lamp, and the shade should be chosen accordingly (see pages 60 to 61 and page 65). Be sure that the shade material will blend with the other furnishings in the room.

In addition to the traditional designs, there are highly individual table lamps in the form of wrought iron sculptures, giant upright eggs, symbolic insects, or robotic arms. Think hard about the value of such a lamp before you buy. Will it remain appealing long enough to make the purchase worthwhile, or is it destined to end up in a dark closet?

Floor Lamps

The large variety of floor lamp designs offers a wide range of lighting options. Floor lamps perform the same function as pendants without the problems posed by permanent installation. Carefully selected and properly used, they can provide a more flexible lighting system than permanent fixtures.

The most common design is the conventional floor lamp, which functions much like a table lamp on a leg. When equipped with three-way sockets, a floor lamp offers a choice of several light levels. It can be used as a reading lamp or it can provide area lighting. Again, the shade must be chosen carefully.

Flexible or jointed models, such as the pharmacy lamp, are particularly good for reading or for close work such as knitting or crocheting. Multiple-source or pole lamps may be freestanding or may extend between the floor and the ceiling. These vertical poles have several adjustable directional fixtures that you can position to light your work.

Tungsten-halogen bulbs are occasionally used in freestanding floor lamps meant for task lighting. One common design based on motion picture studio lamps uses a low-voltage halogen lamp head at the end of a long boom or rod. The transformer on the other end of the rod acts as a counterbalance, and one easy adjustment places the light exactly where you need it.

Freestanding uplights or torchères bounce light off the ceiling. They provide indirect general room illumination without glare. Since the upper edge of the unit should be above one's eye level when one is standing, they are most effective in rooms with high ceilings, which should be white or very light colored. Most uplights stand 6 to 6½ feet high. They will create a hot spot on a standard 8-foot ceiling unless they are used with a dimmer or some form of diffusion.

Specialty Lamps

A specialty lamp is designed to fill a specific need. New styles are always appearing; take your choice. Carefully selected to blend with your decor, specialty lamps provide task lighting at a reasonable cost.

Clip-ons can be used for reading in bed, working at a desk, or highlighting collections displayed on shelves. Floor cans will add instant and economical drama when used to uplight an indoor plant or wash a wall. Minireflector spotlights will highlight paintings or sculptures and won't interfere with general room illumination.

Floor lamps should provide useful illumination and enhance decor.

Surface-Mounted Fixtures

A basic part of any home lighting scheme, fixtures that are permanently attached to the ceiling or walls provide general room illumination. Some can be installed with a minimum of carpentry and rewiring; others require extensive work. Replacing an existing fixture is usually easiest of all.

Manufacturers of surface-mounted fixtures tend to use basic designs that can be installed in various ways. This means that you can light one room or an entire house with a single family of fixtures if you want to maintain decorative continuity.

Ceiling Fixtures

Ceiling-mounted units include fluorescents, ceiling fans with lights, and simple globes. They are usually placed in the center of a given space and used as the general light source. They often provide overall illumination in kitchens, bathrooms, playrooms, hallways, and entries. An important factor when choosing a ceiling fixture is the amount and direction of light that will reflect off the ceiling. If the fixture is attached to a ceiling fan, available headroom is also a consideration.

Wall Fixtures

Wall lights require neither headroom, shelf space, nor floor space. However, they must be placed carefully if they are to do their job. When they are used for general lighting, place them high enough to bounce light off the ceiling. Be aware of potential hot spots when you install one near a corner. In bathroom installations try to achieve a shadowless light for the mirror by mounting a fixture on each side at about the same height as your face or a little lower.

Installing new wall or ceiling fixtures generally involves considerable carpentry and wiring. Special drops will have to be chased inside the wall or across the ceiling to connect to the wiring circuit. If you're not handy with tools, you might want to have these fixtures professionally installed. Be sure to include the cost in your estimated expense.

Pendants

A pendant hung from the ceiling can provide general or task lighting. Since they are conspicuous, pendants should be carefully selected for both function and appearance. Some designs produce only a downward light; others direct light up as well as down. Designs that incorporate plastic or paper globes distribute light uniformly in all directions.

The height of the light source above the floor, the type of shade used, and the size of the room are all important factors in selecting a pendant. For example, an opaque shade produces a cone of light. The lower the pendant, the smaller the light cone it produces. The higher the pendant, the larger the cone.

Consider the size of the room in relation to the light cone. The walls in a small room will reflect some of the light from the pendant, increasing the overall level of illumination. However, there will be dark areas around the edges of a large room if the light cone does not reach the walls.

Size is doubly important when a pendant is used over a dining table. Not only should the fixture visually fit the room, but it should physically fit the table. The bottom of the pendant should be about 30 inches above the tabletop to prevent glare. It should be set in about 6 inches from all four sides of the table so that people standing up to leave the table will not bump their heads.

Adjustable height pendants are useful over dining tables. They can be lowered to change the mood for dining and then raised for general illumination during the evening. These fixtures also make bulb changing easier and less hazardous.

Pendants are often placed in the center of the room to provide general lighting. Unless there are other light sources in the room, this placement lacks flexibility and offers few opportunities to vary the lighting for mood. Consider placing the pendant off center as a

Proper positioning of a light fixture can help to prevent glare. A pendant should be positioned about 30 inches above the surface of a dining table. This will minimize direct and reflected glare.

Proper Positioning of Fixtures

Proper Chandelier Placement

Minimum 6"

30"

means of delineating space, or use several pendants in a large room to accent the furnishings. Using a pendant with matching wall lights can help to relate the different parts of the room to one another by means of a recurring motif.

Chandeliers

Considered by many to be the crowning jewel of a tastefully decorated room, a chandelier is the center of attention, whether on or off. The term *chandelier* encompasses a wide variety of pendant fixtures, from antique reproductions to classic contemporary models, all designed to simulate the hanging candle holders that were used before the advent of electricity.

Chandeliers may be simple and inexpensive fixtures or complex confections of glass and brass adorned with sparkling cut crystal pendants. Some chandeliers even incorporate a hidden downlight to illuminate a centerpiece. Since a chandelier tends to dominate a room, you should seriously consider how best to integrate one into your lighting scheme before you commit yourself. Installing a dimmer to control the light level and establish mood is an excellent idea. The total power of the bulbs should be kept between 200 and 300 watts. When it is not hung over a table, the width of a chandelier in inches can equal the diagonal of the room in feet.

Chandeliers installed above a dining table are subject to the same rules as pendants. In addition, if the height of the dining room ceiling exceeds 8 feet, the chandelier should be raised 3 inches for each additional foot of ceiling height.

A chandelier positioned off center above a buffet that stands against the wall can make a small dining room appear larger. Because they are large and visually dominant, however, many chandeliers look best when centered in a fairly big room.

Recessed Fixtures

A recessed fixture provides illumination without intruding into the room visually and is particularly useful in rooms with low ceilings. Easy to install during new construction or remodeling, recessed fixtures can be added to an existing room as long as there is sufficient space between the ceiling and the floor or roof above.

Installation in an existing room, however, involves considerable expense and effort. A hole must be cut; wiring must be run from a power source to the hole; and the area around the fixture must be touched up after installation. Since they are permanent, recessed fixtures are more suitable for owners than for renters.

Downlights

Cylindrical in shape, downlights contain a lamp in the top half of the fixture and a reflector in the bottom half. They are usually installed flush with the ceiling, although some are designed to project slightly down into the room.

Open downlights are especially useful above work areas in kitchens, where they will spread a strong light on counters and sinks. Hallways, entrances, and stairways also lend themselves to the use of open downlights.

Glare is an ever-present problem with a downlight, but there are several solutions. A common one is to baffle the inside of the fixture with black grooves, reducing the apparent brightness whenever the fixture is in the line of sight. Another is to use a black mirror reflector inside the fixture,

Selecting a Lampshade

Replacing a lampshade sounds easy enough, but simply walking into a store and picking out a new one isn't the way to do it. Chances are good that even if you know the dimensions of the old shade, a replacement bought in this way will not fit properly. Because manufacturers use different frame systems, two shades that look alike may sit differently on the lamp. Slight differences in positioning, shape, or material of the shade can change the look of your lamp, and not always for the better.

Shades are made in three standard shapes or styles. Empire shades have sloping sides, and there is at least a 2-inch difference between the top and bottom diameters. If the difference between the top and bottom is very pronounced, the shade may be called a cone or an umbrella.

Drum shades appear to be cylindrical, but the top diameter is actually 1 inch smaller than the bottom. The depth of the shade determines whether it is a regular, a shallow, or a deep drum.

Bell shades have curved sides. These are produced by means of wire ribs bent inward between the top and bottom rings. Bell shades are always covered with fabric, since paper and other stiff materials cannot be shaped to follow the curve of the ribs.

You can also find specialty shades in a wide variety of shapes and styles. These include squares, ovals, hexagons, pagodas, and Tiffany domes, as well as many variations on the standard shapes: square and cut-corner square bells, scalloped bells, and even hexagonal or octagonal bells.

Regardless of shape, shade size is designated by the bottom diameter. An 18-inch shade measures 18 inches across the bottom. Oval shades are measured across the longer diameter, and hexagonal shades are measured from point to point. Shade depth is the vertical or top-to-bottom measurement. When a shade is described by three dimensions, such as 9 inches × 12 inches × 8 inches, it means that the shade is 9 inches across the top, 12 inches across the bottom, and 8 inches deep.

Shade depth should be about 2 inches less than the height of the lamp base. Deeper shades give the lamp a top-heavy look. Shallower shades often produce a better visual effect, especially if the base is slender. When seen at eye level, the bottom of the shade should be even with the bottom of the socket or harp. If the socket shows, either the shade is too short or the harp is too deep. If the problem is the harp, it can be replaced. Replacement harps come in 7-inch to 13-inch sizes.

The replacement shade should be appropriate to the base. Shade designers recommend a plain shade with a decorated base, a textured shade with pottery, and an exquisitely designed shade for a porcelain base.

Function is just as important as appearance. The shade should work together with the lamp. Light-colored, translucent shades provide diffuse light in addition to the direct light emitted through their top and bottom openings. Dark or opaque shades reduce the level of illumination produced by the lamp.

When you have bought a shade and installed it on your lamp, don't forget to remove the cellophane wrapper. If you leave it on, the heat given off by the bulb will gradually dry out the cellophane, causing it to tighten. This can compress and permanently warp the shade frame.

# OW TO INSTALL LIGHTING

If you've read this far, you've created a lighting plan, and learned about the various light sources and fixtures that are necessary to make it work. Now that you understand the basics, it's time to put everything together and improve your home lighting. In this chapter you'll learn about basic tools and how to work with wiring to install wall receptacles, switches, and light fixtures. You'll discover how to map circuits and determine load capacities. As you progress through the chapter, you'll learn to recognize the jobs that you can do and the ones that require the services of an electrician.

With the table reflecting warmth and the candles adding their glow, this dining room would serve perfectly for an intimate dinner for two or a dinner party for a small group of friends.

BEFORE YOU BEGIN

You will have to decide how much electrical work you are comfortable doing yourself. Be sure and follow the commonsense safety precautions, and do not overload any circuits.

Should You Do It Yourself?

You can do the electrical work necessary to install your own lighting. Once you understand how to replace different types of switches and receptacles, you will be able to add a few more lights in a room or hallway and even to extend an existing circuit. For more involved projects refer to Ortho's book *Basic Wiring Techniques.*

How much and what kinds of wiring you can do yourself depends on your local building department and its regulations. Although it's legal to do your own electrical work as long as you have the required permits, the red tape and delays connected with some jobs may make it more practical to let a professional do it.

If you are working on an older home that has not been rewired, you may encounter porcelain knob-and-tube wiring or aluminum wiring. These are quite different from, and incompatible with, modern nonmetallic cable installations. Even if local regulations don't prevent you from dealing with this situation, you may decide that now is the time to have the old wiring replaced.

Discretion is the better part of valor. If you are not completely comfortable doing the job, find yourself getting in over your head with problems you don't understand, or have doubts about the best way to proceed, it's time to seek the help of an experienced electrician.

Safety

Electricity is dangerous only if you don't respect it. Whenever you are working with or around electricity, there are prescribed safety measures that you should observe. By following some commonsense rules, you'll avoid any risk to yourself while you are doing the work, and any risk to those who will be using the system afterward.

The basic rule is this: *Always shut off electrical power to the area in which you are working.* Then make sure that it is off by using a voltage tester before you touch any wires. If there is no electricity in the wires, you cannot be hurt. Here's a checklist.

• Never work on a live circuit. Before starting, remove the fuse or turn off the breaker that controls the circuit in question. Put the fuse in your pocket or tape the breaker in the *off* position. Post a sign on the box to warn others not to touch it because a circuit is being worked on.

• Before touching any wires, use a voltage tester to make sure that they are dead. To test a receptacle, insert one probe into each slot; if the lamp on the tester lights up, the circuit is still live. To test a switch, remove the face plate, touch one probe to the metal box and the other probe first to one screw terminal and then to the other. If the box is plastic, touch one probe to the screw terminal and the other to the bare grounding wire. If the lamp lights up when you touch either screw terminal, the power to the circuit is still on.

• Never stand on a damp floor when working with electricity. The combination of moisture and electricity can be lethal. If the floor around the fuse or circuit breaker box is even occasionally wet, put down some boards or a rubber mat to stand on before touching the box.

• Never touch any plumbing pipes or gas pipes when working with electricity. They are often used as the ground to the electrical system. Touching a hot wire and the pipe at the same time could send the current through your body.

• Don't use aluminum ladders near overhead entrance wires.

• After completing the electrical work, turn the power on and check it with the voltage tester. The tester should light when a connection is made between the hot black wire and the grounded box. It should not light when a connection is made between the white wire and the grounded box.

Wiring Codes and Regulations

The *National Electrical Code* (*NEC*) is a body of rules and regulations that specify the methods and materials to be used in safe electrical work. Although the *NEC* is applied uniformly across the United States, local and county codes often overlap or amend it and thus may be more demanding. For example, the *NEC* approves of aluminum wiring, but many local codes now forbid it. You must abide by the code that has jurisdiction over your residence. If you live in a city, follow the local codes; if you live in an unincorporated area, follow the county codes.

Since both local and county codes are at least loosely based on the *NEC*, you need to be familiar with it, too. Don't bother obtaining the code itself

Wire Color Coding

Color	Function
Black	Hot wire
Red	Hot wire
Blue	Hot wire
White coded black*	Hot wire*
White	Neutral wire
Green	Grounding wire
Green and yellow	Grounding wire
Bare copper	Grounding wire

* White is always a neutral wire, with the exception of a switch loop (described on page 95), where it must be identified as hot with a dab of black paint or wrapping at the end with black electrician's tape.

Wire Sizes, Ampacity, and Use

Number	Ampacity	Use
No. 4/0	195	Service entrance wire
No. 3/0	165	Service entrance wire
No. 2/0	145	Service entrance wire
No. 1/0	125	Service entrance wire
No. 1	110	Service entrance, grounding wire
No. 2	95	Service entrance, grounding wire
No. 4	70	Individual appliances (240 volts), grounding wire
No. 6	55	Individual appliances (240 volts), grounding wire
No. 8	40	Individual appliances (120 volts)
No. 10	30	Individual appliances (120 volts)
No. 12	20	Small appliance and lighting circuits
No. 14	15	Lighting circuits
No. 16	10	Doorbells
No. 18	7	Flexible cords, low-voltage systems

Note: Outlets and switches have ampere ratings that are to be matched to the type of wire being used. Most of them are stamped "15 amp," which means they should be used with No. 14 wire. However, the *NEC* permits you to use 15-amp switches and outlets with No. 12 wire, which has an ampacity of 20 amperes.

and trying to read it. It's complex and unenlightening. Instead, buy a paperback guide to the code. When you have read it, talk with the electrical inspector in the local building department about the requirements for permits and inspections. Obtain a copy of the local code. Find out what you can do yourself, what must be done by a professional, and what steps you must follow to ensure that the work will be approved.

Generally you can replace receptacles and switches without a permit, but a permit is required to extend an existing lighting circuit or install a new one. Some areas prohibit homeowners from doing their own wiring. Others will let the homeowner install the new circuit, but it must be connected to the service panel by a licensed electrician. The law in some jurisdictions may restrict the type of lighting that can be installed in specific areas. For example, energy-conscious areas often insist on the use of fluorescent lighting in kitchens and bathrooms.

The *NEC* sets minimum standards, but there is nothing to prevent you from installing a circuit that exceeds the specified minimum. In fact, when undertaking a major project, such as rewiring a house, adding a circuit, or even extending one, always check with your fire insurance company before choosing materials or deciding who will do the work. Some companies offer lower insurance rates when the new work exceeds the minimum standards required by the *NEC*. For example, they may offer a lower rate for wiring with armored cable, even though the code allows the use of nonmetallic cable. The rate may also be lower if a licensed electrician does the job.

If you need a permit, submit a short written description of the project, complete with a list of wire sizes and details on all other materials, such as fuses, breaker switches, light switches, and receptacles. Include a simple floor plan or schematic drawing showing where the wire will run and where the junction boxes, switches, and receptacles will be located. It need not be elaborate as long as it is clear. The inspector will then go over these plans and indicate any changes that may be required to meet local codes. Once the plans have been approved, you will receive a permit, for a fee. When the project is completed, the inspector will check the work and give the final approval.

Determining Electrical Needs

The electric company sends power to your home through an underground cable or overhead wires. The power goes first to the company's meter, where it is measured and the amount used is recorded. The incoming electricity then goes to the service panel, where it is distributed to the individual circuits. These individual circuits are protected by fuses or circuit breakers in the service panel and carry power to various points in the house.

A *circuit* is the path electrical current takes from the service panel through an electrical device and back to the panel. A house with modern wiring (nonmetallic cable) contains three kinds of circuits: general-purpose, or lighting, circuits; small-appliance circuits; and individual-appliance circuits. Before you do any work on the circuitry, you need to understand how the house is wired.

Home Wiring

Electricity is brought to the house in either two or three wires. If the house was built before 1940 and the wiring has not been modernized, there will be only two wires to the meter box. Two-wire electrical service does not allow the use of 240-volt appliances.

Each of the two wires carries 120 volts. One is a *hot* wire, which means that it carries ungrounded electricity. The other is a *neutral* wire, which completes the electrical circuit. Although this second wire is technically neutral because it is grounded, it still carries current and must be handled carefully in case it is hot. Hot wires are normally black or red and neutral wires are normally white. (Exceptions are explained at appropriate points in the chapter.) However, never assume that the color actually indicates whether a given wire is hot. Trust only the voltage tester. Treat all wires with healthy respect.

In modern houses there are three wires to the meter box, supplying both 120 and 240 volts. Two of these three wires are hot and the third is neutral. The two incoming 120-volt wires are connected to a bus bar inside the service panel.

Single circuit breakers that connect to one end of the bus bar will provide a 120-volt circuit. When a double circuit breaker is installed, however, it picks up power from both ends of the bus bar to provide the 240-volt circuits required for such major appliances as a range or a clothes dryer.

Electrical current is also measured in amperes. Amperage is limited by the size of the wire through which it passes. If the current sent through a wire is too great, the wire will overheat. This causes the fuse to blow or the breaker to trip, shutting off the flow of current before a fire can start. General-

purpose, or lighting, circuits use 15-ampere or 20-ampere wiring and fuses.

The incoming electricity is distributed to the individual circuits in the house by the service panel. You can disconnect all power to the house by throwing a switch, pulling down a lever, or pulling a main fuse block at the service entrance.

If the house has a circuit breaker panel, the main disconnect switch is normally located at the top of it, just above the individual circuit breakers. If the house has a fuse box, the main power can be disconnected in one of two ways.

Either the box has a lever at the side, which is pulled down, or it has two cartridge fuses in a block, which is pulled out. Both fuses and circuit breakers allow you to disconnect the main power safely without touching any live wires.

The fuses or breakers controlling the circuit should be numbered to correspond with numbers on a list inside the service panel door. This list should identify the items controlled by each circuit— "kitchen range," "air conditioning," "living room lighting," and so on. It should also identify the circuit amperage.

Circuit Plan for a Typical Home

Lighting and Outlet Circuits

— Kitchen light, porch light, family room, and upstairs bath

— Small-appliance circuit

— Small-appliance circuit

— Dining room, downstairs hall, and two upstairs bedrooms

— Upstairs hall, linen closet, downstairs bath, entrance-hall light and outlets, and porch light

— Living room and master bedroom

— Garage, workshop, and outside light

Individual-Appliance Circuits

Electric clothes dryer

Range and oven

Air conditioner

Dishwasher

Garbage disposer

⊙ Light

s Switch

● Outlet

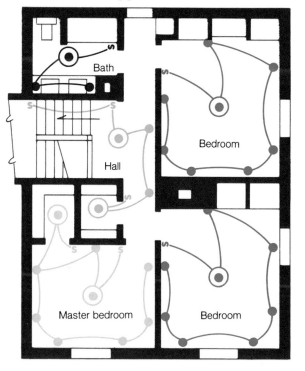

Upper Level

Bath

Hall

Bedroom

Master bedroom

Bedroom

Identifying the Circuits

Before you install any new lighting, know what each circuit currently controls. Some circuits may already carry the maximum amount of current for which they are designed.

To identify the circuits and to determine what each one controls, start at the service panel. Copy the circuit numbers from the list inside the panel door onto a separate piece of paper. If there is no list, consecutively number each fuse or breaker switch in the panel and copy those numbers onto the paper.

Next draw a floor plan of the house. Include the basement, the garage, and the attic (if it is wired for electricity). Walk through each room and mark the approximate location of each fixture, appliance, receptacle, and switch on the floor plan.

Now turn on all the lights and switches in the house. Turn on any radios, TVs, or other small appliances and make sure that all lamps are plugged into their receptacles. Do not turn on major electrical appliances, such as stoves or dryers, since they are on 240-volt circuits.

Remove the first fuse or turn off the first circuit breaker. Go through the house and check all the appliances, receptacles, and switches. Use a night-light or small lamp to check any outlets not in use. Indicate all appliances and outlets that are dead by writing the circuit number beside their location on the floor plan.

If you live in a two-story house, make sure to check both stories every time you break a circuit. It is common for one circuit to go to both floors. At least two circuits generally serve each story to prevent a blown fuse or breaker from darkening an entire floor.

Return to the service panel and reinstall the fuse or turn the circuit breaker back on.

Remove the next fuse or turn off the next circuit breaker and repeat the procedure. Continue in this manner until you have identified all the circuits in the house and what they control.

When you have finished, the floor plan will serve as a circuit map. This map will help you to locate the trouble area quickly when a fuse blows or a breaker trips. Furthermore, if you need to cut power to one part of the house in order to make repairs, you will know immediately which fuse or breaker will do the job. If there was no circuit list inside the service panel door, use the map to prepare one and post it.

Lower Level

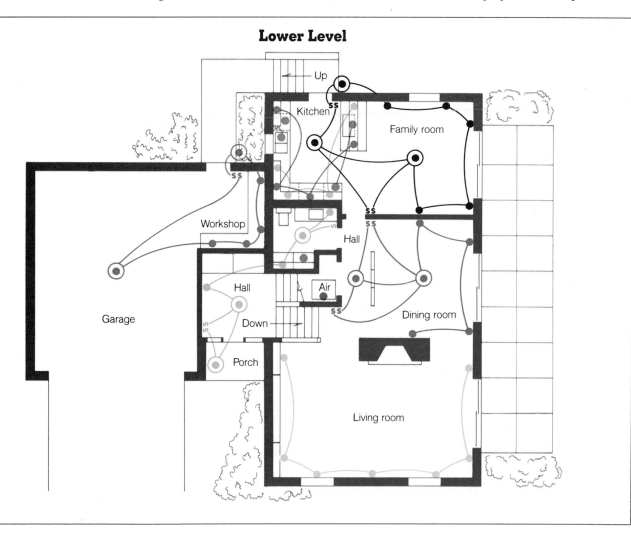

Calculating Load Capacities

After you have plotted the circuits on a floor plan, you can determine the rating of each circuit and the load that each one currently carries. This will tell you which lighting circuits can take additional fixtures or receptacles. Be sure that you know exactly what the local code specifies; adding lights to a 20-ampere circuit is illegal in some areas.

The maximum capacity of a circuit is given in watts. It is determined by multiplying the amperage by the voltage. For example, a 15-ampere rating on a 120-volt circuit permits a maximum of 1800 watts (15×120) to be used on that circuit. Anything over 1800 watts will overload the wires and cause the fuse or circuit breaker to blow.

In practice it is just as well to underrate the circuit somewhat to accommodate the occasional use of an additional light or appliance. Assuming a maximum of 1650 watts on a 120-volt circuit with a 15-ampere rating provides a safe margin. With a 20-ampere rating, do not exceed 2200 watts.

To calculate the load on a given circuit, add up all the watts used by the lights and appliances that are served by that circuit and that are commonly in use at the same time. (Watt ratings are printed on light bulbs and on a plate attached to any appliance or power tool.) Subtract the wattage in use from 1650 watts.

This is the total number of watts that you can safely add to a 15-ampere circuit. If the number of watts you want to add exceeds this total, you should either reduce the wattage or run a new circuit.

Basic Tools

You don't need many tools even for extensive wiring projects, and those you do need aren't expensive. You can do a great deal with a screwdriver, a hammer, a pocketknife, and a pair of pliers.

With a few of the more specialized tools described in this section, you can tackle almost any home wiring job. Consider tools an investment and always buy the best quality you can afford. Cheap tools are generally a false bargain; they break quickly and have to be replaced.

Wire strippers make wiring jobs easier and faster. It is possible to strip and cut wire with a knife and a pair of pliers, but a stripper saves a lot of time and frustration. The multipurpose or combination stripper tool is the best investment, since it can cut and strip all common wire sizes and also clamp terminals in place.

Lineman's pliers are a basic electrical tool. A good pair has serrated jaws to hold the wire and wire cutters located behind the jaws. Diagonal cutting pliers can be substituted for lineman's pliers.

Needle-nose pliers are useful to hold or bend wires. Two or three different pairs with various nose lengths will let you choose the right tool for the job.

Cable rippers are used to remove the outer layer of flat two-wire nonmetallic cable. After using the ripper, you can easily remove the insulation.

A voltage tester is a miniature electrical circuit. It has no power of its own. Power enters through one wire of the tester, travels through the bulb, and goes out the other wire. It is used to find hot wires, to determine whether receptacles and switches are live or dead, and to check for proper grounding. Since the circuit being tested must be on, be sure to hold the tester probes by their insulated ends and avoid touching the metal probe tips.

A continuity tester is used to determine whether a circuit is complete, or continuous, without having to turn on the house current. If the circuit is complete, the small battery in the tester will light up the bulb. If the circuit has been interrupted, the bulb will not light. Because it contains its own power system, never use a continuity tester where electrical current is flowing.

Fish tape is a long, flat piece of spring steel or stiff nylon that can be pushed through small holes and down into wall spaces. One end is bent into a hook and is used to catch and hold the hooked end of new wiring that is pushed through holes drilled behind the walls. Fish tape comes in reels up to 50 feet long for easy handling and compact storage. When a wire must be pulled only a short distance, a straightened coat hanger or a length of No. 12 wire makes an adequate substitute.

Flat-blade and Phillips screwdrivers are basic tools. It is best to have a set that includes an assortment of common heads and sizes. If you are buying a set, look for electrician's screwdrivers. These are longer than ordinary screwdrivers, and have insulation extending down the stem almost to the tip. If you can't get these, wrap the stems of regular screwdrivers with electrician's tape. The added insulation can be useful when working in service panels or outlet boxes. It prevents the user from grounding the screwdriver to the metal edge of the panel or box.

A solder gun is used to splice tie-ins when working with the old style knob-and-tube wiring.

For more comprehensive wiring work, you will also need a variety of common hand tools. These include a cold chisel, a wood chisel, an electric drill with bit extensions and a variety of wood and masonry bits, hacksaws, and hammers.

Basic Wiring Tools

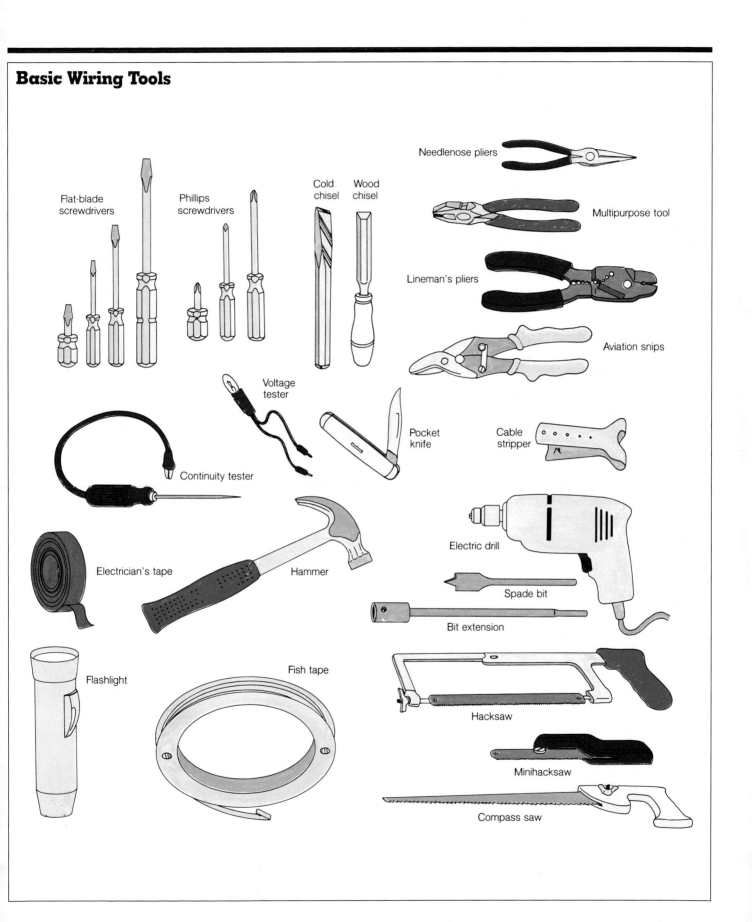

Flat-blade screwdrivers

Phillips screwdrivers

Cold chisel

Wood chisel

Needlenose pliers

Multipurpose tool

Lineman's pliers

Aviation snips

Voltage tester

Pocket knife

Cable stripper

Continuity tester

Electrician's tape

Hammer

Electric drill

Spade bit

Bit extension

Flashlight

Fish tape

Hacksaw

Minihacksaw

Compass saw

CABLE AND WIRING

All the components of an electrical circuit are interconnected by some form of cable or wire. Cable is used for all the fixed wiring in a circuit. Pendants and portable fixtures, such as table lamps and floor lamps, are connected to the circuit with a length of flexible cord.

Types of Wire and Cable

Technically a *wire* is an individual strand; a *cable* is two or more wires combined in the same sheathing. Each wire in a cable is individually insulated to keep the wires from touching and creating a short circuit.

Type T wire is very common in household installations. The wire is wrapped in thermoplastic (T) insulation as protection against a wide range of temperatures, from below freezing to around 150° F.

NM cable is the commonest form of nonmetallic cable. It is widely known by the trade name Romex. NM cable consists of two or more individually insulated Type T wires and a bare copper grounding wire, all enclosed in a plastic sheath. The space between wires is filled with jute, which can act like a wick to attract moisture. This wire must not be used in damp locations.

NMC cable is a nonmetallic cable that is specially designed to be used wherever moisture may be present, as in basements or laundry rooms. NMC cable often has a plastic wrapping on each wire. The wires are embedded in a solid plastic sheath to keep out moisture.

Armored cable is commonly known by its trade name, BX. Usually made of flexible steel or aluminum, the spiral armor itself acts as a partial ground, aided by a strip of aluminum. The wires inside are wrapped in heavy paper to protect them from the abrasive action of the cable. Armored cable cannot be used in damp locations and must be used only with metal boxes.

Conduit is a protective metal tube for wiring. It is specified by some local codes under certain circumstances. The conduit is installed first, and the wiring is drawn through it. The conduit also acts as a ground for the wire.

If your house was constructed between 1964 and 1973, the wiring may be aluminum rather than copper. If it is, be sure to read the sections on receptacles and switches before working on it. (See page 91 and page 93.) There are two types of aluminum wire: all-aluminum and copper-clad aluminum. All-aluminum wire is now considered a possible fire hazard and is prohibited by some local codes. If you are at all unsure what type wiring you have, consult an electrician.

Sizes and Uses of Wire

The size of the wire is important for the efficient flow of electricity. The more current the wire must carry, the larger the wire must be to accommodate the flow. Too much current in a wire will cause it to overheat and damage the insulation or even cause a fire. Therefore the NEC stipulates the type and size of wire that must be used in all wiring projects.

Wire sizes are designated by the American Wire Gauge (AWG) system. The size of the wire is commonly stamped on the insulation. The wire type (such as NM for nonmetallic cable) is stamped beside the wire size. In wire sizes the higher the number, the smaller the wire. Thus, No. 12 wire is larger than No. 14 wire. (See the wiring chart on page 79.)

Most house wiring for a lighting circuit uses No. 12 or No. 14 copper wire. No. 12 wire is preferable, although No. 14 is considered to be standard. If you are installing new wiring, it is advisable to use No. 12, which is more efficient at carrying current than No. 14.

Receptacles and switches are rated according to their ampere capacity and should be matched to the type of wire being used. Most are stamped 15 amps, which means that they should be used with No. 14 wire. However, the NEC permits the use of 15-ampere switches and receptacles with No. 12 wire, even though No. 12 wire has a 20-ampere capacity. Note that the circuit should be fused for 15 amperes despite the higher capacity of the wire.

The small wires, such as No. 16 (10 amperes) and No. 18 (7 amperes), are used for systems with low-voltage requirements, such as doorbells, and are stranded rather than solid for greater flexibility. They are also used for light fittings and lamp cords. Small wires used for these latter purposes come in two types, flat and round.

The flat, or zip, cord is insulated with a molded thermoplastic and can be separated into two individual sheaths, each covering a hot wire and a neutral wire. The round cord consists of two wires (each covered with insulating material) encased in a plastic covering.

Wires are wrapped in a colored insulation that designates their function and helps to prevent errors in connecting them. Neutral wires are white. Grounding wires may be green or green and yellow, or they may be left bare. Hot wires are usually black but may be red or blue. However, never rely solely on the color code to identify hot wires. Always use a voltage tester to determine whether a wire is hot.

Outlet Boxes

An outlet (also called a housing box) provides a connection point inside a wall or ceiling for joining wires or for installing light fixtures, switches, or receptacles. In addition to housing the electrical components, the box keeps the exposed ends of the wire away from flammable material.

Boxes are made either of steel with a galvanized finish or of plastic. Plastic boxes cost less than metal ones and are satisfactory for installing a basic lighting circuit or for modernizing old work. Because plastic boxes do not conduct electricity, they need not be connected to the grounding

Sizes and Types of Wire and Cable

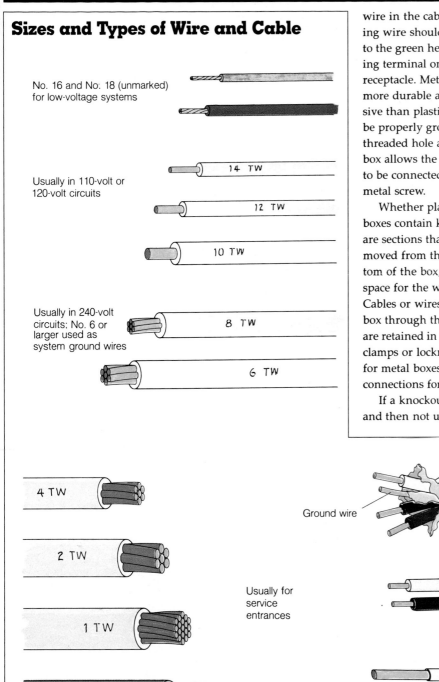

No. 16 and No. 18 (unmarked) for low-voltage systems

Usually in 110-volt or 120-volt circuits

14 TW

12 TW

10 TW

Usually in 240-volt circuits; No. 6 or larger used as system ground wires

8 TW

6 TW

wire in the cable. The grounding wire should be connected to the green hexagonal grounding terminal on the switch or receptacle. Metal boxes are more durable and more expensive than plastic, but they must be properly grounded. A threaded hole at the rear of the box allows the grounding wire to be connected with a sheet-metal screw.

Whether plastic or metal, all boxes contain knockouts. These are sections that can be removed from the sides or bottom of the box, providing a space for the wire to enter. Cables or wires entering the box through these knockouts are retained in place by built-in clamps or locknut connections for metal boxes or by snap-in connections for plastic boxes.

If a knockout is removed and then not used, the code specifies that the unused hole must be sealed with a knockout closure. This can be either a metal disc with tension clips around the edge or two discs placed one on each side of the hole and held together by a screw through their centers.

Ceiling Boxes

Ceiling boxes can be square, octagonal, or round. They are used both to protect the wiring connections of a ceiling fixture and to provide a mounting point for the fixture. They may be flanged for fastening to the side of a joist, or they may be attached to a hanger bar for installation between two joists. Ceiling boxes are also used in either walls or ceilings to enclose cable connections or spliced wires.

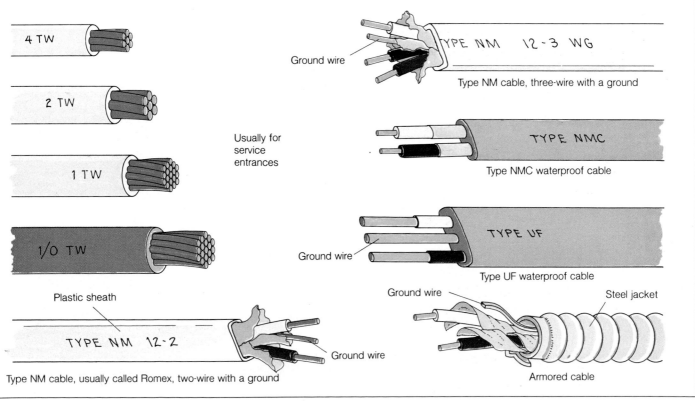

4 TW

2 TW

1 TW

1/0 TW

Plastic sheath

TYPE NM 12-2

Type NM cable, usually called Romex, two-wire with a ground

Ground wire

Usually for service entrances

Ground wire

TYPE NM 12-3 WG

Type NM cable, three-wire with a ground

TYPE NMC

Type NMC waterproof cable

Ground wire

TYPE UF

Type UF waterproof cable

Ground wire

Steel jacket

Armored cable

Wall Boxes

Wall boxes are designed to hold either outlet receptacles or switches. They are always rectangular. The most common type of wall box is installed flush with the surface of a wallboard or lath-and-plaster wall. A hole of the appropriate size is cut, and the box is inserted and fastened to the surface of the wall with screws.

Some metal wall boxes have a mounting flange for attaching them to the front, rear, or side of a joist or stud. Plastic boxes can be attached with a special bracket in much the same way.

Metal boxes with spring-ear clamps (often called cut-in boxes) are designed to be used in areas between studs. Once installed with the clamps, they cannot be removed without damaging the surface of the wall. Before installing a box of this type, check its fit without the clamps and make sure that the cables are in place.

Some wall boxes can be ganged, or joined, by removing one side of each box and then screwing the two boxes together. This is useful when more space is needed than one box can provide.

Stripping Wire

Before cables or wires can be connected, their covering must be removed to expose the bare copper. To remove plastic or rubber sheathing, lay the wire flat on a smooth surface. Six inches back from the end of the wire, insert the tip of a sharp knife into the sheathing and make a shallow cut down the center to the end. Using this cut as a guide, make successive passes with the knife. Cut just deep enough to penetrate

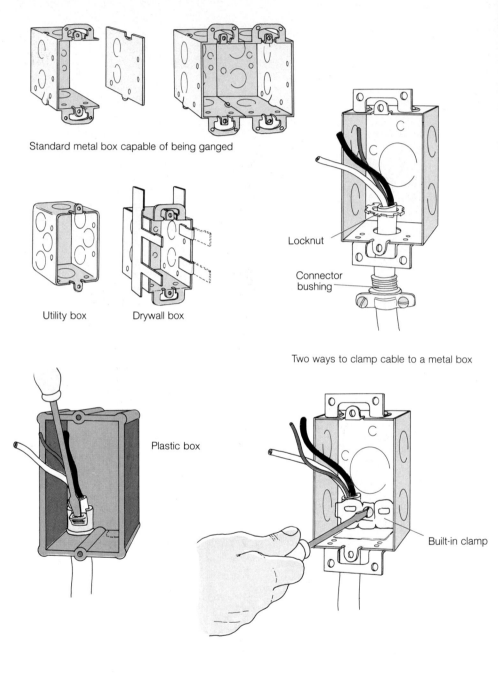

Wall Boxes and Connectors

Standard metal box capable of being ganged

Utility box   Drywall box

Locknut

Connector bushing

Two ways to clamp cable to a metal box

Plastic box

Built-in clamp

sheathing without cutting the insulation around the wires. Once the sheathing is split, peel it back and cut it off with a knife or a pair of wire cutters.

Stripping metal requires a hacksaw. To remove armored

cable, make a cut about 8 inches from the end, so that you will have plenty to wire for the connections. Hold the saw at a right angle to the spiral strip of the armor rather

than at a right angle to the cable itself. Cut only through the armor; do not cut the aluminum grounding strip or the wires inside. This is not easy; it takes practice. Once you have made a cut through the top of the armor, give the cable a sharp bend at the cut point and twist it back and forth to break the rest of the armor.

The next step is to strip the insulation. It is best not to use a knife, because you may cut into the wire and weaken it. If you have nothing but a knife, don't cut at a right angle to the wire; you are more likely to damage the wire that way. Hold the knife blade at about a 60-degree angle. Then twirl the wire back and forth against the blade until you have cut through the insulation all around the wire. Pull the insulation off with your fingers.

Using a wire stripper or a multipurpose tool makes the job of wire stripping much faster and easier. Simply place the wire in the correct hole (they are numbered by wire size), squeeze the handles, and rotate the tool to break the insulation or sheathing. Pull the tool off the wire and the covering will come off with it.

Splicing Wire

Electrical cable cannot be connected or spliced in a run; connections must be made only inside outlet boxes. When wires are joined together, the connection must be very tight. A poor connection can cause a drop in voltage. It can also cause a fire, set either by overheated wires or by the spark that occurs when the electricity tries to jump between loose wires. Wires that are stripped and then spliced together must be fully insulated again. Most connections today are made with wire nuts or crimp connectors. In older homes that still have knob-and-tube wiring, soldered splices must be

used to bring current into an outlet box.

Wire Nuts

A wire nut is a solderless connector consisting of a plastic insulating shell with a threaded, tapered, copper interior. Wire nuts come packaged in different sizes corresponding to wire sizes. A chart on the package will tell you which size to use.

To make a wire nut splice, first remove 1 inch of insulation from the ends of the wires to be connected. Hold the stripped ends together and twist them clockwise one or two turns. Then cut off about ½ inch of the twisted wires to even the ends. Insert the twisted wires into the nut and turn the nut clockwise to screw it down as tightly as possible. To ensure that the nut will not come loose, wrap the base and the wires with a few turns of plastic electrician's tape.

Crimp Connectors

Crimps can be used to join No. 16 or No. 18 wire. First, strip about ½ inch of insulation from the wires. Place them in the crimp connector and squeeze it together with a multipurpose tool or crimping pliers. Tug lightly on the wires to make sure that they are tight. Do not use crimps with aluminum wire; it expands and contracts and will eventually work loose.

Solder Splices

Soldering is required only to connect new wiring to the hot and neutral wires in a knob-and-tube circuit. All wires to be soldered must be made of cop-

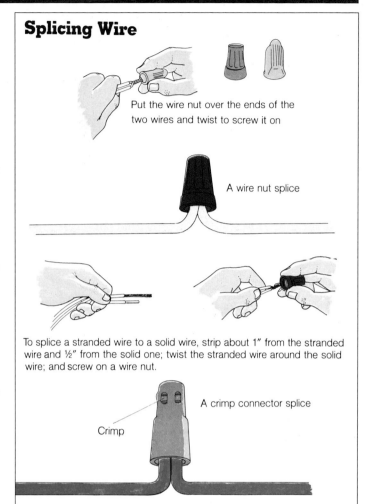

Splicing Wire

Put the wire nut over the ends of the two wires and twist to screw it on

A wire nut splice

To splice a stranded wire to a solid wire, strip about 1" from the stranded wire and ½" from the solid one; twist the stranded wire around the solid wire; and screw on a wire nut.

A crimp connector splice

Crimp

per. Rosin core solder should always be used. The wires must be absolutely clean or the solder will not adhere.

With the circuit off, strip about 2 inches of insulation from the old wiring and about 1½ inches from the new connecting wire. Clean the ends of both wires with a piece of sandpaper or emery paper until they are shiny. Wrap the end of the connecting wire around the old wire tightly to form a coil about ¾ inch long. Then cut off the protruding ends.

Heat the coil for a few seconds with the tip of a 250- or 300-watt solder gun. Holding the gun in place, touch the end of a roll of rosin core solder to

the connection. If the coil is hot enough, the solder will melt from the heat of the connection. If it does not melt, remove the solder and continue heating the connection with the gun. When the solder melts, let it flow through and around the connection before removing the tip of the gun from the connection. Remove the tip carefully so that it will not disturb the connection before the solder has cooled.

Once the solder hardens and the coil cools, wrap the soldered connection with rubber insulation tape. Secure the tape in place by wrapping it in the opposite direction with plastic electrician's tape.

EXTENDING A CIRCUIT

If you want to install additional fixtures in a room, it may be necessary to add switches and receptacles by extending an existing circuit. This involves tying into the circuit at a certain point, determining where the new components are to be added, then running the wire through the wall or under the floor to that location.

Selecting the Circuit

You must tie into the correct type of circuit, as specified by the code. The first step is to make sure that adding the desired fixtures, switches, or receptacles will not overload the circuit. To do this, check to see whether the fuse or breaker switch is 15 or 20 amperes. With a 15-ampere circuit, a safe maximum total load is 1650 watts (see page 76). With a 20-ampere circuit, total wattage should not exceed 2200. Before extending a 20-ampere circuit, be sure that the local code permits the addition. As a rule of thumb, the total number of fixtures, switches, and outlets on one circuit, including any additions, should not exceed 12.

Choosing the Tie-in Point

The next step is to determine the best point to tie into the existing circuit. First, find the easiest route to run the new wiring. This will depend somewhat on how the house is constructed. The easiest route may be through a wall, under the floor, or above the ceiling.

Once you have determined an approximate path for running the wiring, look for the best tie-in point. You can tie into a circuit at almost any point where there is a junction box, fixture, switch, or receptacle. The exceptions are a switch-controlled fixture or receptacle at the end of the circuit and a switch box that contains no neutral wire.

An end-of-the-run receptacle that is not switch controlled is usually the easiest place to tie into an existing circuit. Since the box has only three wires from one cable, the wiring is easier to handle. If the outlet is controlled by a switch, however, the circuit extension will have power only when the switch is on.

A middle-of-the-run receptacle or switch has two cables coming into the box. If you install a third cable to extend the circuit, you will have to gang two boxes together to make enough room for the additional wires.

To tie into an existing circuit, the box must be able to accommodate the new wires. It must have a knockout hole for running the new cable into the box. If the location is good but the box does not meet these criteria, replace it with a box that does.

Installing a Junction Box

Junction boxes are commonly used to extend a circuit or to split one incoming power source into two or more separate circuits. The code requires that a junction box always be visible, accessible, and fitted with a plain cover plate to

protect the wiring inside. For this reason junction boxes are installed in basements, attics, and closets, but never sealed behind a wall.

To add a junction box to an existing circuit, first locate an accessible cable. With the power off, cut the cable at the place where you will install the junction box. This should be at some point along a stud or over a joist. Strip the wires at each end of the cut cable and at the end of the new cable that will provide the extension.

Next, remove the knockouts from a metal junction box for the incoming and outgoing original wires plus the new wires. Attach the box to the stud or joist and then slip the three cables into the box under the clamp connectors. Tighten the screw on each clamp to force it against the cable, securing the cable in place.

If the box uses a locknut connector instead of a clamp, slip a bushing over each cable and tighten the two screws to

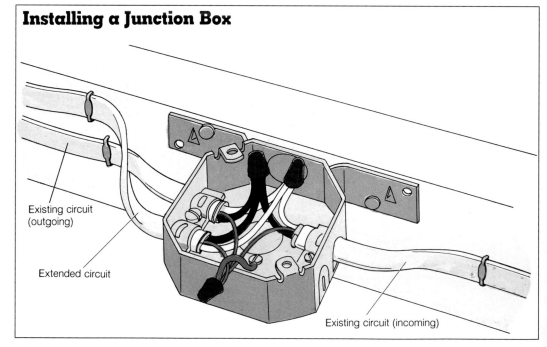

Installing a Junction Box

Existing circuit (outgoing)

Extended circuit

Existing circuit (incoming)

Tying Into a Ceiling Fixture

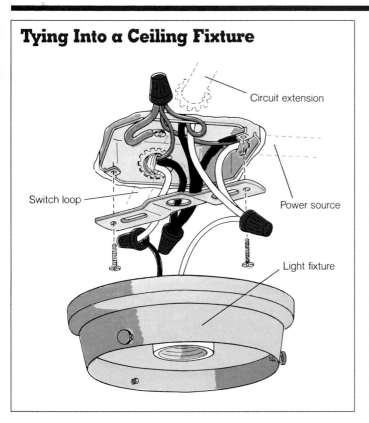

Circuit extension

Switch loop

Power source

Light fixture

clamp it in place before inserting the cable into the box. Push the cable and bushing into the box through the knockout hole. Thread the locknut onto the bushing and tighten it by turning its serrated edges with a suitable screwdriver blade.

Join the three black wires together and install a wire nut; do the same with the three white wires. Join the three bare grounding wires with a pigtail and install a wire nut; then connect the other end of the pigtail to the metal box with a sheet-metal screw.

Tapping Into a Ceiling Fixture

A ceiling fixture is another convenient place to tap for power when you want to extend a circuit. The new cable can be run from the ceiling through a wall to new outlets. The ceiling

fixture takes on the added function of a junction box.

With the power off, open the ceiling fixture to expose the wiring. If the ceiling fixture has only two wires coming into it, it is not an appropriate candidate for tapping. The two wires indicate that it is at the end of a switch-controlled run and will have no power when the switch is off.

You will find an incoming white wire attached to the fixture, and an incoming black wire connected to the white one that leads to the switch. This indicates a switch loop, and the white wire should be painted partly black or have black tape on it to indicate that it is hot. The bare grounding wires are attached to each other with a pigtail, which is attached to the metal screw on the fixture box.

To extend the circuit, cut the white wire leading to the fixture. Use a wire nut to connect the cut wires to the white wire on the circuit extension. Connect the circuit extension black wire to the connection between the incoming black wire and the hot white wire (marked black) that leads to the switch. Finally, join the circuit extension grounding wire to the other grounding wires.

Tapping Into Receptacles

If you have a lot of plug-in lamps and not enough outlets, or if you want an outlet in a particular place, you can extend the circuit from an existing receptacle by running new wiring across the wall behind a baseboard and up to the new outlet.

If the receptacle to be tapped is at the end of a run, it will contain only three wires (black, white, and grounding). To extend the circuit from this receptacle, connect the new black wire to the brass colored screw of the box and the new white wire to the silver colored screw. Connect the grounding wire to the existing grounding wire with a wire nut. Power arriving at this receptacle will pass through it and continue along the new wiring to the added receptacles.

A middle-of-the-run receptacle contains six wires (two sets each of black, white, and grounding). A bigger box will be required to hold an additional set of three wires. You can gang two boxes together as described on page 80 to provide room for the extra wires. To tap into this receptacle, use wire

nuts to connect black to black, white to white, and grounding to grounding, including a pigtail with each connection. (Cut short lengths of each wire and strip their ends to make the pigtails.) Connect the other ends of the pigtails to the appropriate screws or push-in connections in the box. The pigtails will connect the incoming power to both the existing receptacle and the circuit extension.

A wall switch is another convenient tap point for extending a circuit. Since switches are generally installed about 4 feet from the floor, the circuit extension will run down through the wall and across to the place where the new outlets are installed.

Before shutting off power to the switch, use a voltage tester to find out which of the two black wires coming into the box is hot. Now shut off the power and disconnect that black wire from the switch. Cut a short pigtail and strip the ends. Connect one end of the pigtail to the switch. Using a wire nut, connect the other end to the incoming hot wire and to the black wire of the circuit extension. Connect the white and grounding wires in the circuit extension to the existing white and grounding wires with wire nuts.

Installing New Boxes

Before tapping the power source and running the circuit extension cable, you should determine how many and what kind of boxes you will need and then mark and cut the appropriate holes in the wall or ceiling.

Connecting a New Wire to an Existing Circuit

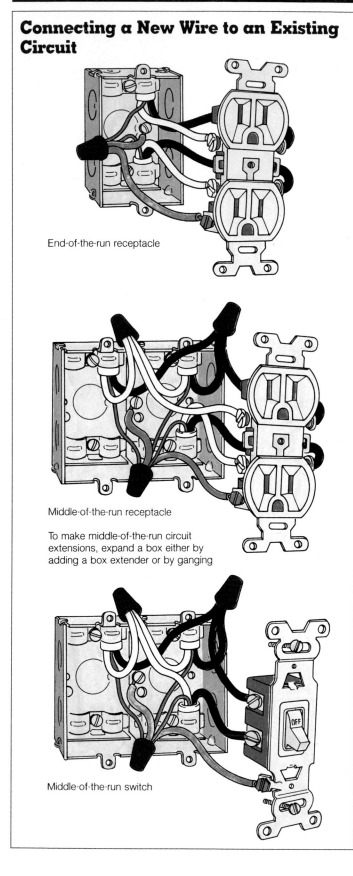

End-of-the-run receptacle

Middle-of-the-run receptacle

To make middle-of-the-run circuit extensions, expand a box either by adding a box extender or by ganging

Middle-of-the-run switch

Plain or cut-in boxes are satisfactory for switches, outlets, and fixtures that weigh less than 20 ounces. These can be installed in the space between studs or joists. For heavier fixtures use a flanged box or one with a hanger and secure it to the studs or joists.

Locating the Opening

If you plan to mount a box on a stud behind the wall, or in the wall between the studs, you must locate the stud first. There are three ways to do this.

Rap on the wall sharply with your knuckle. Go back and forth until you hear a solid, as opposed to a hollow, sound. The solid sound indicates the presence of a stud.

Use a stud finder. Its magnetic needle will swing into line when it passes over one of the nails that hold the wall sheathing to the stud.

Drill a ⅛-inch test hole in the spot where you plan to put the box. If it is over a stud, wood shavings from the stud will come out of the hole.

Before drilling any holes, shut off the power to any circuit that runs behind the wall (or ceiling). Also locate any possible obstructions behind the wall, such as plumbing.

To mount the box between the studs, drill a test hole first. Take a piece of stiff wire 8 to 10 inches long, bend it in half at a right angle, and insert one end into the hole. Rotate the wire in all directions to make sure that there are no obstructions. If you do run into an obstruction, move over a few inches and try again. Insulation in the walls can make this operation somewhat difficult.

Follow the same procedure for lath-and-plaster walls. Then chip away enough plaster around the hole to expose the nearest lath. The box should be exactly centered over this lath. Use a cold chisel to chip away plaster until the opening is centered accordingly. Work carefully in order not to crack the surrounding plaster.

Cutting the Opening

Hold the front of the box against the wall (or ceiling) and trace its outline. Do not trace the outline of the brackets, if there are any.

Drill a pilot hole at each corner of the outline. Then use a saber saw or a keyhole saw to cut along the outline; remove the cutout section. With plaster walls, apply strips of masking tape on the outside edge of the outline before you cut. This prevents the plaster from cracking. If you are cutting a hole in a plaster ceiling, use one hand to brace the ceiling as you cut.

Routing New Cable

Once you have cut the holes for all the new boxes, run cable from a power source to each hole. Running new cable in the walls of a finished house can be fairly simple or quite tedious, depending on how and where it is routed. In any case, the job will take two people.

One difficulty is that the wall studs, through which the wiring is normally run, are covered. The easiest way to avoid this problem is to run the new wiring under the house or through the attic. If that is not possible, the wire can be run behind the baseboard, or even through the studs, by removing a section of the wallboard. However, walls covered with materials such as ceramic tile that are difficult to patch should not be broken into.

You will need a few special tools for running additional wire. These include fish tape (see page 76) and an electric drill with several different bits. A $\frac{1}{16}$-inch bit is used for drilling pilot holes and a ¾-inch space bit for drilling holes through joists and plates. An 18-inch extension bit often comes in handy, too.

The code requires that cable be fastened within 12 inches of a new outlet or switch and every 4 feet along joists or studs. Although rustproof staples are permitted by code, using straps with two nails will prevent any damage to the cable from a slightly misplaced staple.

Behind a Baseboard

It is often quite easy to add one or more receptacles along a wall by removing the baseboard and running the wire behind it. This is particularly true in lath-and-plaster walls, where the baseboards may be 8 inches high. The following method can be used in either wallboard or lath-and-plaster.

1. Use a stiff putty knife to score through the layer of paint between the wall and the baseboard. Then carefully pry the board away from the wall with a thin chisel. If you pull on one end of the loose baseboard instead of using the chisel, you may crack it.
2. With the current off, remove the receptacle from the existing box, leaving the wires intact.

Using a screwdriver, punch out the bottom knockout from the existing box to accept the new cable.
3. Cut a hole in the wall for the new box.
4. In lath-and-plaster, chisel a channel in the plaster to the depth of the studs between the existing box and the hole cut for the new box, along the midline of the baseboard. In wallboard, remove a 2-inch-wide section along the line of the baseboard between the old and the new boxes.
5. Drill a ⅜-inch hole in the wall at each end of the channel beneath the two boxes. Fish the cable up to the existing box and connect the wires.

6. Run the cable down the channel and fish that end up to the new box. Enclose the cable in metal raceway to protect it from nail punctures.
7. Insert the cable into the new box and secure it in place with the clamp or locknut connector. Install the new box in the hole. Then connect the wires and install the receptacle.
8. Replace the baseboard. Be careful not to puncture the raceway.

Behind Wallboard

The walls in most modern houses are made of 4 by 8 wallboard sheets. Here is the easiest way to run new cable

behind this type of wall.
1. With the current off, remove the receptacle from the existing box, leaving the wires intact. Using a screwdriver, punch out the bottom knockout from the existing box to make room for the new cable.
2. Starting from the center of the stud that holds the existing receptacle and proceeding across the wall to the place where the new outlet will be located, measure, cut and remove a 6 inch-wide piece of wallboard. Stop in the middle of the stud where the new box will be attached. This step must be performed precisely so that the replacement piece can be fitted smoothly into the opening.

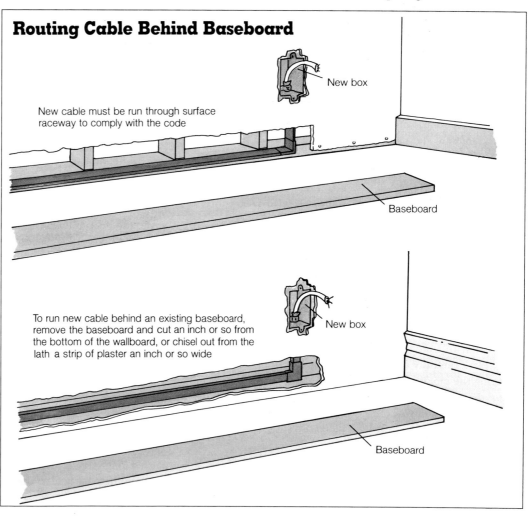

Routing Cable Behind Baseboard

New cable must be run through surface raceway to comply with the code

New box

Baseboard

To run new cable behind an existing baseboard, remove the baseboard and cut an inch or so from the bottom of the wallboard, or chisel out from the lath a strip of plaster an inch or so wide

New box

Baseboard

INSTALLING FIXTURES AND RECEPTACLES

Once wiring has been run, installing a receptacle is just a matter of connecting up wires . . . the correct way, of course. Fixtures may be just as simple, or may require substantial structural work.

Surface-Mounted Fixtures

Surface-mounted fixtures are installed either on a wall or on the ceiling. Although they come in a wide variety of designs, they are all supported in one of three ways. They may be attached directly to the outlet box, to a mounting strap, or to a threaded stud.

Mounting Methods

The mounting hardware provided with new fixtures may differ according to the manufacturer and the type of light, but it can generally be adapted to any standard fixture box. Many small fixtures are designed to attach directly to the ears of the box. These fixtures have screw terminals for attaching the cable wires.

Some wall fixtures are designed to be mounted on studs with cap nuts. These fixtures take a cross-shaped mounting strap. One leg of the strap screws directly to the outlet box; the other leg contains the studs for mounting the fixture.

Larger fixtures are supported from a central threaded stud, or are screwed to a mounting strap, or both. If there is a central stud in the ceiling box that does not fit the threaded nipple on a new fixture, a reducing nut from a lighting shop will correct the problem. This nut has different threads at each end to fit both the stud and the nipple.

If the new fixture is designed to screw directly to the ceiling box, but the screw holes do not align, get a mounting strap sized to fit inside the cover plate on the fixture. Screw the strap to the ceiling box and then attach the fixture to the strap.

Hooking Up Ceiling Fixtures

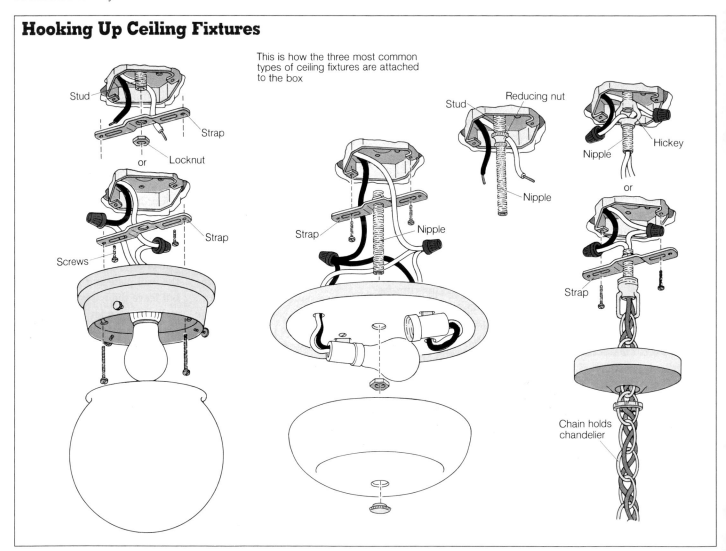

This is how the three most common types of ceiling fixtures are attached to the box

Stud — Strap — or — Locknut — Strap — Screws

Strap — Nipple

Stud — Reducing nut — Nipple — Nipple — Hickey — or — Strap — Chain holds chandelier

An alternative method is to screw the strap to the ceiling box and then connect the fixture to the strap with a threaded nipple through the hole in the center of the strap. If the fixture wires must be run through the nipple to the fixture, use a hickey, as shown on page 83.

A fixture that weighs over 50 pounds cannot safely be attached to a ceiling box. It should be attached to a joist or beam as well as to the box.

Grounding the Fixture

All light fixtures with exposed metal must be grounded by code. If the box used with the fixture is grounded, attaching the fixture to the box will usually solve the problem. Fixtures suspended from the box by a cord or chain, however, must have a grounding wire between the box and the fixture. New fixtures are usually provided with a grounding wire.

If your house is an older one, the wiring may not be grounded. If it is not, ground the box to a cold-water pipe before you attach the fixture. Install a ground strap on the nearest pipe. Wrap one end of a length of bare No. 12 copper wire to the fastener holding the fixture to the box. Route the wire as unobtrusively as possible to the ground strap and connect it to the strap screw.

Replacing a Fixture

One inexpensive and painless way to give rooms a fresh look is to install new ceiling lights. Before you dismantle the existing fixture, turn off the circuit breaker or remove the fuse that

controls that light. Just turning off the light switch is not sufficient protection; some of the wires in the box may still be live.

First, carefully remove the fitting cover, or on a chandelier or ceiling fan-light, lower the canopy. Figuring out how to do this can sometimes be baffling. There may be several screws around the edge or a single, disguised cap nut in the center. If there is neither, try turning the cover plate to the left to disengage it.

Next, remove the fitting from its bracket or unscrew the hickey from the ceiling box stud. Once you have removed the fixture, don't let it hang by the wires. Bend a wire coat hanger and connect one end to the ceiling box; then connect the other end to the fixture. This will support the fixture so that you can use both hands to pull the wires out of the box and unscrew the wire nuts. On older fixtures the wires may be spliced with electrician's tape instead of wire nuts. If they are, unwrap and discard the tape. When the fixture is disconnected, carefully remove it from the coat hanger and place it safely out of the way.

Since most ceiling lights are located at the end of either a circuit or a T-branch, there should be wires from only one cable in the box. If there are more, either the circuit continues on or a switch loop has been added. In either case note carefully how the wires are connected before you disconnect the fixture.

To install the replacement fixture, strip about ½ inch of the insulation from its wires and follow the steps outlined above, in reverse order.

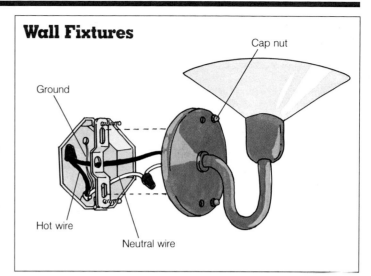

Wall Fixtures

Cap nut

Ground

Hot wire

Neutral wire

Installing a New Fixture

Once you have routed the cable from a power source and installed the switch and fixture box, the installation of a new fixture is essentially the same as the replacement of an old one. (Complete procedures for routing the cable and installing the switch are outlined on pages 84 to 87 and pages 93 to 97.) Here's a quick recap of the necessary steps.

1. Decide where to put the light. Trace the outline of the ceiling box and cut out the marked hole.

2. Cut a hole 4 inches wide at the point where the ceiling meets the wall, to expose the top plate. This hole should be directly in line with the hole for the ceiling box.

3. Decide where to put the switch. Cut a hole for the switch box between 44 and 48 inches from the floor and about 15 inches from the door opening. Keep the switch at the same height as any other switches in the room.

4. Run the power cable from the nearest existing outlet to a point directly below the hole for the switch. Feed a fish tape from the switch hole down to the cable. Connect the cable to the tape and pull the cable up and out through the switch hole, allowing about 12 inches of the cable to protrude from the hole.

5. Continue the switch loop from the switch hole to the hole at the point where the ceiling meets the wall. Fish the switch loop cable up to this hole and over to the ceiling box hole. Again, let about 12 inches of cable protrude from the ceiling box hole.

6. Notch the double top plate so the cable can pass through it into the ceiling cavity. Clamp or staple the cable in place. Cover the cable with metal protector plates.

7. Clamp the cable in each box, strip the insulation, and connect the wires to the existing outlet and the switch.

8. Install the ceiling box with the cable clamped into it. Install the fixture, using wire nuts on all splices.

9. In wallboard walls and ceilings, patch the holes with pieces of wallboard, leaving a ⅛-inch crack on all sides. Press wallboard joint compound into the cracks and apply strips of perforated joint tape, with butt (not overlapping) corners. Press the tape into the compound and smooth it out with the putty knife. Spread a thin, smooth layer of compound over the tape. Let it dry and then apply three more coats. Sand the final coat smooth and paint it to match the wall or ceiling.

In lath-and-plaster, patch the holes with plaster patching compound to about ⅛ inch below the surface of the wall. Let it dry and then smooth on a coat of patching compound. Let this dry, sand it smooth, and paint it to match the wall or ceiling.

Installing Recessed Lighting

Installing more than one recessed light can be a chore. Most incandescent downlights are circular cans that must be wired into a junction box mounted on a joist. Recessed fluorescent fixtures are often prewired and grounded to an attached outlet box.

In either case there must be sufficient clearance above the finished ceiling to accept the fixture. For this reason recessed fixtures are best installed in a ceiling that has an unfinished attic or a crawl space above it.

Recessed fixtures must be mounted so that there is no combustible material within ½ inch of the fixture. If there is insulation above the ceiling, remove enough of it to leave at

least a 3-inch space all around. Following these precautions and using the correct bulb in the fixture will prevent fires.

Cut the ceiling hole following the standard procedure. Locate the joists; decide where to place the light; trace the outline of the fixture on the ceiling; cut it out with a saber saw or a keyhole saw.

If the fixture does not have its own box, fasten a junction box to a joist near the hole. Run the power cable to the box. Then clamp the cable of the fixture to the box and connect the wires. The metal-clad cable will ground the fixture to the box. Connect the fixture housing to its socket, position the assembled fixture in the hole, and insert the clips that hold it to the ceiling.

Two types of fixtures come with premounted boxes. One is a complete unit with adjustable hanger bars. This is ideal for installation from above. After positioning the assembly in the hole, secure the hanger bars to the joists. Run the power source cable into the fixture box and make the connections. The fixture trim or baffle plate is installed from the ceiling side of the installation.

The other type of fixture can be disassembled from the mounting frame that contains the outlet box. This fixture is installed from below. Choose it when there is no access above the ceiling. The mounting frame and box can be angled through the fixture hole in the ceiling, positioned, and clipped to the edge of the hole. Make the circuit connections and snap the fixture into its socket and into the mounting frame.

Installing a Track System

Track systems are surface mounted on the wall or ceiling, and fixtures can be installed at any point on the track. The system can be attached with mounting clips, suspended, or recessed. Recessing a track system, however, is recommended only with new construction. Cutting up the walls and ceiling to recess the track in an existing room is a major project. It must be done by a professional if it is to look right.

Like metal raceways, tracks receive power from either plug-in connectors or wire-in connectors. When low-voltage fixtures are used, they should have integral transformers. If they do not, plug-in adapters can be used. If the system requires an external transformer to step down power to the track, wire must be routed to the track from the transformer.

Connectors and Switches

Power is usually supplied to a track system by a connector installed at one end of the run. Some manufacturers offer an in-line connector designed for permanent mounting that will bring in power at any point along the track. A plug-in connector has a length of cord with a plug extending from one end. A wire-in connector has two wires coming out of the top.

Plug-in connectors must be installed on the surface of a wall or ceiling. The track is positioned near an outlet and connected to it with the cord. Plug-in connectors can be used only with single-circuit tracks.

Wire-in connectors are designed to tap directly into an outlet box. They can be surface mounted or recessed. A special saddle or cover is provided for the fixture box. The saddle positions the unit and covers the wiring connections.

When replacing a single fixture with a single circuit track, you can use the wall switch for the old fixture. A track system that is installed to supplement other lighting in the room will require a new wall switch for each of its circuits.

Mounting Methods

Tracks are mounted in different ways. Most are held in place with screws or toggle bolts. Tracks that use a wire-in connector are sometimes installed with special clips that position the track a fraction of an inch from the surface of the wall or ceiling. It is easier to install tracks if you have an assistant.

Using a chalk line or a yardstick and pencil and starting from the centerline of the connector, lay out the routing of the track. There are knockout holes in the roof of each section; position the first section on the line and mark the location of the knockout holes for drilling. Remove the track and drill the holes. If the track mounts directly to the surface, have your assistant attach it to the connector and hold it in place while you fasten it with the screws or toggle bolts. If clips are required, attach them to the surface and slip the track onto the connector, pressing it into the clips. Install the other sections of track in the same way. If the last section is too long for the run, it can be cut with a hacksaw.

Wiring Plug-in Outlets

Receptacles are generally installed 12 to 18 inches from the floor, or within easy reach of small children. If there are small children in your house, consider installing safety outlets. These outlets are fitted with solid covers that must be rotated before a plug can be inserted. At the minimum, you should use plastic caps. These are designed to fit flush into the socket when it is not in use. You can pry them out easily; small children cannot.

Reading a Receptacle

All new outlets should be stamped UND. LAB. This indicates that the outlet meets the rigid safety standards of the Underwriters' Laboratories, Inc. Outlets are rated and marked for a specific amperage and voltage; be sure to buy the correct type for the installation. In addition, a series of letters stamped on the mounting yoke indicates the type of wire to be used. Aluminum wire requires the use of receptacles with special terminal screws. These receptacles are marked CO/ALR. Receptacles marked CU or CU CLAD are for use only with copper or copper-clad aluminum wire. Be sure to match your outlets to your cable.

Types of Receptacle

There are three basic types of receptacle: side wired, back wired, and combination.

The side-wired receptacle has four terminal screws: two brass colored and two silver colored. The hot wires, black or red, go on the brass colored screws; the white neutral wires go on the silver colored screws. The green hex screw at the bottom is for attaching the grounding wire.

The back-wired or push-in receptacle has holes in which to insert the wires. A strip gauge on the back of the outlet shows how much insulation to strip off the ends. The word *white* is stamped on one side of the plug, indicating that the white neutral wires go in there. The hot, red or black, wires go in the other side. To release the wires, push a screwdriver into the slot above or below the holes. Aluminum wire *cannot be used* in push-in receptacles.

Combination receptacles have both push-in holes and terminal screws. Either method of connecting the wires is acceptable.

Most receptacles are duplexes, so-called because they contain two units. Both units are hot at the same time because they are joined by a brass tab or ear, which passes the current from one side to the other. On the other side is a chrome or silver tab that keeps current flowing in the white neutral wire. If the brass tab is broken off by twisting it with a pair of pliers, only the outlet that is actually wired will be hot. The current will not pass to the other. This is done when you want to control one half of the outlet with a switch.

Replacing or Installing Receptacles

The diagrams on page 92 show common wiring situations. The first and second figures are for outlets wired in parallel with both halves always hot. In the third figure, two receptacles are installed side by side to provide four outlets. This installation is made by joining two switch boxes together or by using a single 4-inch-square box intended for two devices. In the fourth figure, both halves of the receptacle are controlled by a wall switch. In the fifth figure, the bottom half is always hot, and the top half is controlled by a separate wall switch. The sixth figure shows two different ways to replace a switch with a combination switch-receptacle. In this case it must be middle-of-the-run wiring because an outlet must always be hooked to a white neutral wire to complete the circuit.

Where two wires would be installed on a single screw terminal, the code requires that a wire nut and a pigtail be used instead. Twist the two wires and the pigtail together with a wire nut. Bend a loop in the other end of the pigtail and tighten it down under the terminal screw.

Before starting work, be sure to disconnect the circuit by switching the breaker off or removing the fuse. After connecting the wires, fold them into the box and attach the outlet with its screws, adjusting mounting slots to position the outlet correctly. If the outlet isn't flush, it can be shimmed with washers. The final step is to install the face plate. Then turn the circuit back on and check your work with a night-light or a plug-in lamp.

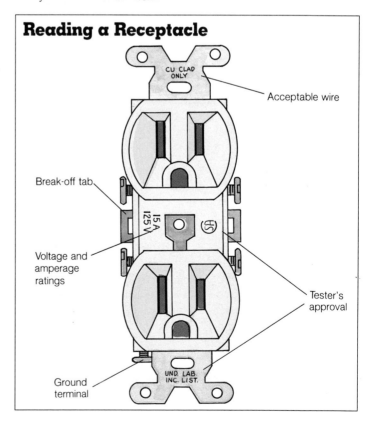

Reading a Receptacle

Acceptable wire

Break-off tab

Voltage and amperage ratings

Tester's approval

Ground terminal

91

Wiring Receptacles

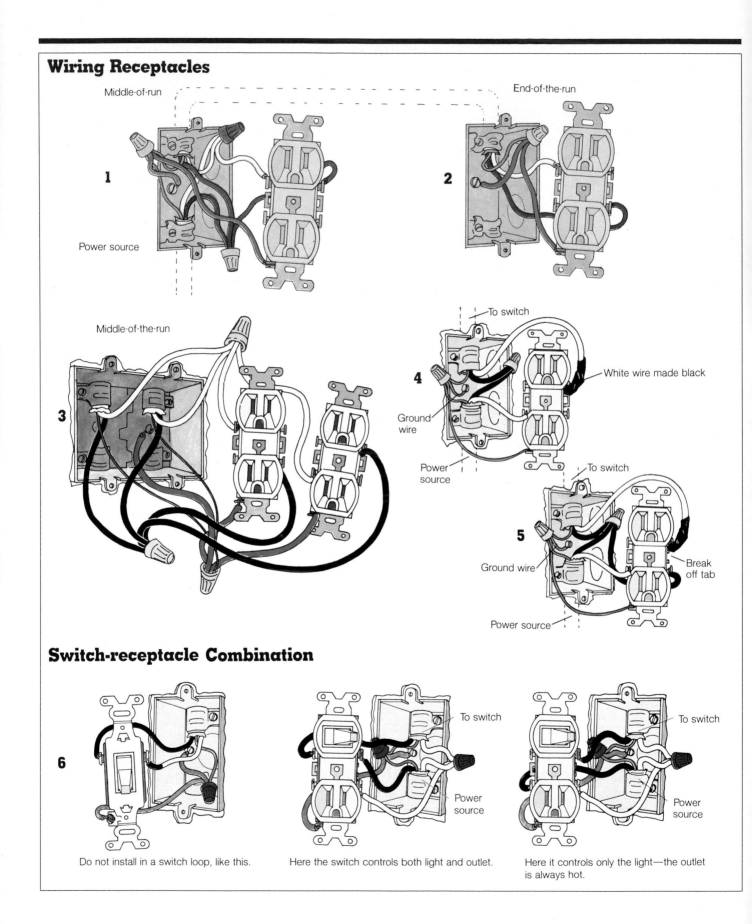

Middle-of-run

End-of-the-run

1

Power source

2

Middle-of-the-run

3

4

To switch

White wire made black

Ground wire

Power source

5

To switch

Ground wire

Break off tab

Power source

Switch-receptacle Combination

6

To switch

Power source

To switch

Power source

Do not install in a switch loop, like this.

Here the switch controls both light and outlet.

Here it controls only the light—the outlet is always hot.

WIRING A SWITCH OR A DIMMER

Most homes use either single-pole or three-way switches. Both control one or more lights or outlets; three-ways are used in pairs to control from different points. Occasionally a four-way switch is installed between two three-way switches. The four-way switch provides lighting control from three or more locations.

Reading Switches

Like receptacles, switches have important information stamped on the metal mounting strap or yoke. Make sure that the switch is stamped UND. LAB. to ensure that it meets safety standards, and that it has the same amperage and voltage rating as the circuit with which it will be used. For lighting circuits with a maximum capacity of 15 amperes, use a switch marked 15 A-120V.

New switches are stamped AC ONLY. This means that they can be used only with

alternating current, the only type now available in residences. In older houses the switches that click loudly when moved were designed for either alternating current or direct current. These switches have no markings and should be replaced with switches marked AC ONLY.

Finally, but equally important, if the house has aluminum wiring make sure that the switch is stamped CO/ALR. If it is marked CU/AL, it can be used only with copper or copper-clad aluminum wire.

Types of Switches

Like receptacles, switches may be side wired, back wired, or both. All three designs work equally well. With the side-wired switch, the cable wires must be wrapped around the screw terminals on the side. Back-wired switches are installed by stripping the ends of the wire and pushing them into the appropriate slots on the rear of the unit. Remember, push-in terminals should not be used with aluminum wire. Combination switches can be connected by means of either the terminal screws or the push-in terminals. (There are also front-wired and end-wired switches, but these are uncommon.)

Replacing or Installing Switches

Switches seldom wear out, but you may want to replace an old one with a modern one or with a different model. If you suspect that a switch has failed,

first check the bulb; then check the circuit fuse or breaker. If both are good, cut the power on the circuit by removing the fuse or turning the breaker off. Then remove the face plate over the switch.

Before removing the switch, use a voltage tester to make sure that it is dead. Probe each terminal screw while the other tester probe touches the metal box (or the bare grounding wire, if the box is made of plastic). If the switch has push-in terminals, probe each release slot in the same way.

If the tester glows at any time, the circuit is still hot and you have removed the wrong fuse or tripped the wrong breaker. Go back and find the proper fuse or breaker for the circuit. If the circuit is dead, remove the top and bottom yoke screws and pull the switch out of the box. Then use a continuity tester to determine whether the switch works properly.

If a three-way switch appears defective, check both switches in the pair as described above to determine which one is malfunctioning.

Single-Pole Switches

Two black wires and a bare grounding wire will be connected to the box. The two white wires will be joined by a wire nut and can be left alone. This arrangement indicates that power goes through the switch box to the light fixture.

Loosen the screws holding the black wires or depress the push-in terminals to disconnect them. Transfer the black wires to the new switch. With a single-pole switch, it does not matter which black wire goes

Reading a Switch

Everything you need to know to choose the proper switch is either stamped into the mounting yoke or molded into the back of the plastic case. Study it all carefully

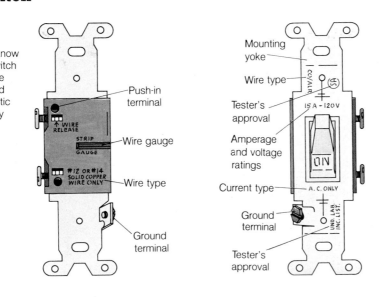

Push-in terminal
Wire gauge
Wire type
Ground terminal

Mounting yoke
Wire type
Tester's approval
Amperage and voltage ratings
Current type
Ground terminal
Tester's approval

How Switches Are Wired

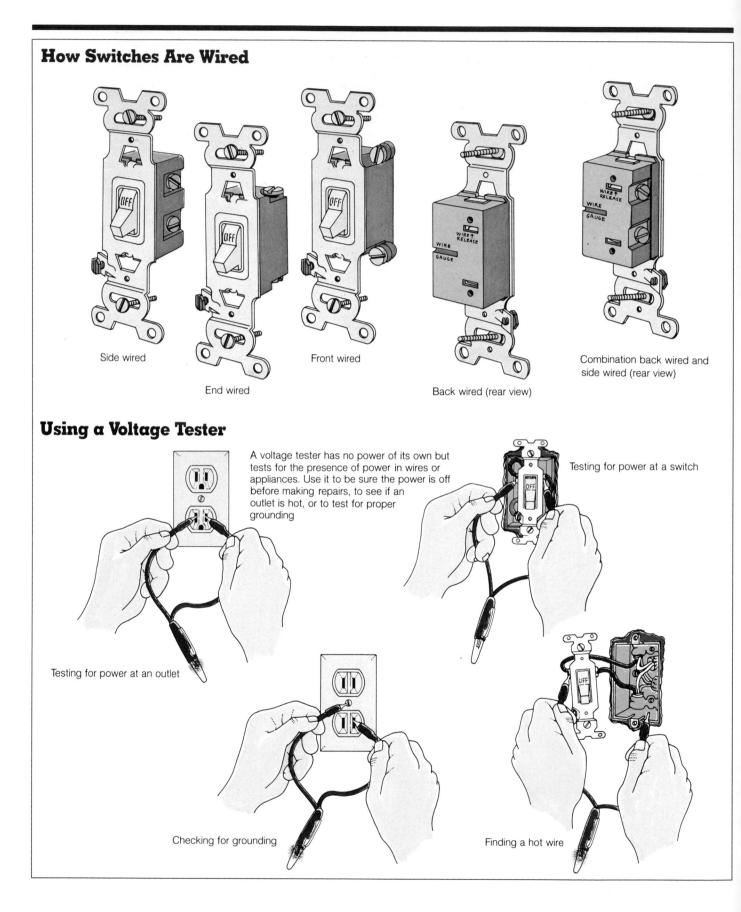

Side wired

End wired

Front wired

Back wired (rear view)

Combination back wired and side wired (rear view)

Using a Voltage Tester

A voltage tester has no power of its own but tests for the presence of power in wires or appliances. Use it to be sure the power is off before making repairs, to see if an outlet is hot, or to test for proper grounding

Testing for power at a switch

Testing for power at an outlet

Checking for grounding

Finding a hot wire

Replacing a Single-Pole Switch

Power Coming From Source

Power Coming From Fixture (switch loop)

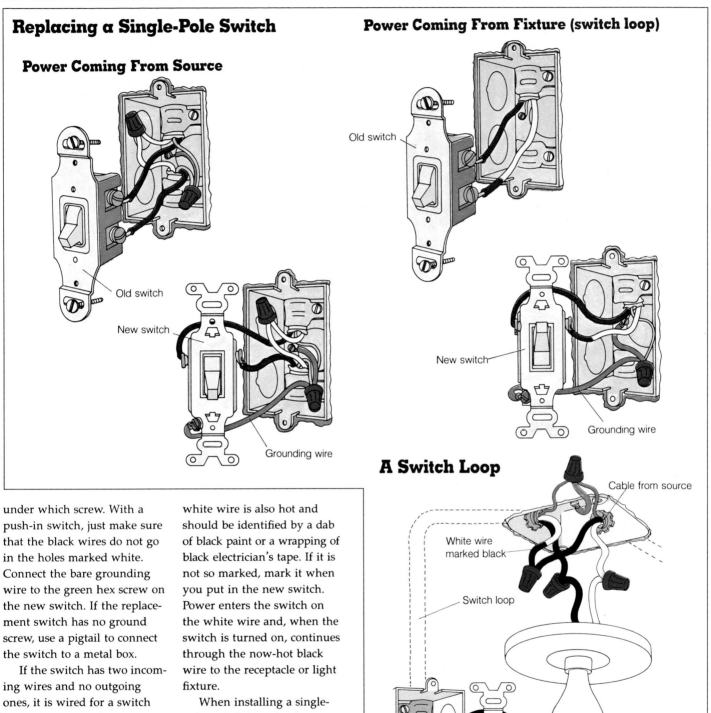

Old switch

New switch

Grounding wire

Old switch

New switch

Grounding wire

A Switch Loop

Cable from source

White wire marked black

Switch loop

White wire marked black

under which screw. With a push-in switch, just make sure that the black wires do not go in the holes marked white. Connect the bare grounding wire to the green hex screw on the new switch. If the replacement switch has no ground screw, use a pigtail to connect the switch to a metal box.

If the switch has two incoming wires and no outgoing ones, it is wired for a switch loop. This means that power goes through the light fixture before arriving at the switch. Although it is the reverse of standard wiring, the switch loop is a common technique for extending an existing circuit. When a switch loop is used, the white wire is also hot and should be identified by a dab of black paint or a wrapping of black electrician's tape. If it is not so marked, mark it when you put in the new switch. Power enters the switch on the white wire and, when the switch is turned on, continues through the now-hot black wire to the receptacle or light fixture.

When installing a single-pole switch in a circuit extension, join the white wires (if any) with a wire nut. Join the grounding wires and a pigtail with a wire nut. Then connect the pigtail to the green hex screw on the switch. Connect the hot wires to the terminals on the switch.

Replacing a Three-Way Switch
Power Coming From Source

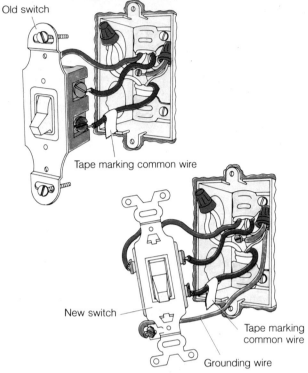

Old switch

Tape marking common wire

New switch

Tape marking common wire

Grounding wire

Power Coming From Fixture

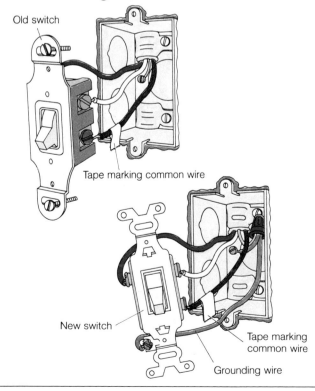

Old switch

Tape marking common wire

New switch

Tape marking common wire

Grounding wire

Three-Way Switches

Before removing any wires from a three-way switch, identify the common terminal. Either it will be marked COMMON or it will have a black or copper colored terminal screw to distinguish it from the other two screws (which are brass and silver colored, respectively). Once you have identified the common terminal, put a piece of tape on the wire connected to it.

Remove the other wires from the switch. These will include a hot black wire, a hot red wire, a white neutral wire, and a green or bare grounding wire. Reconnect these wires in the same way to the new three-way switch and then connect the marked wire to the common terminal.

If the bare grounding wire was not connected to the green hex screw on the old switch, use a pigtail and a wire nut to connect the new switch to the grounding wires. Use another pigtail to connect the new switch to a screw in the metal box as a ground.

When installing a pair of three-way switches in a circuit extension, connect the hot wire from the power source to the common terminal on one switch. Connect the hot wire from the light fixture or receptacle to the common terminal on the other switch. Connect the two remaining terminals on one switch to the terminals on the other switch with hot wires. Use a pigtail and a wire nut to connect the switches to the grounding wires. Use a pigtail to connect the switches to a screw in the metal box.

Four-Way Switches

A four-way switch can be identified by its four terminal screws. All four wires connected to the switch are hot. A red (or possibly black) wire and a white wire marked as black (or possibly a black wire) are connected to the top terminal screws. A second pair of identical wires are connected to the bottom terminal screws. The easiest way to replace the switch is to remove the wires running to the top terminals and put them on the same top terminals of the new switch. Repeat the process with the wires on the bottom terminal screws. Do not disturb the black wires joined by a wire nut in the box. Use a pigtail to connect the grounding wire to the green hex screw on the switch and to the metal box.

When installing a four-way switch in a circuit extension, join the black switch with a wire nut. Connect the grounding wire to the green hex screw on the switch and to the metal box with a pigtail. Connect one hot wire of each color to the top terminals. Then connect the remaining hot wires to the bottom terminals.

Dimmer Switches

A dimmer switch contains a rheostat to reduce the flow of current and so to change the intensity of the light. The dimmer switch makes it possible to create different lighting effects, saves energy, and prolongs the life of the bulb.

Dimmers are normally used to control overhead incandescent lights. Special dimmers are required for fluorescent tubes and for low-voltage halogen bulbs. To use a dimmer with fluorescents, you will need rapid-start tubes and a special two-wire dimming ballast. The amount of wattage that a dimmer can handle is stamped on the switch. Knob dimmers can usually accommodate up to 600 watts and toggle switch dimmers up to 300 watts. Be sure to use a dimmer switch that is appropriate for the fixture.

Wall-mounted dimmer switches are available in single-pole and three-way designs. When a dimmer is used in a three-way switch, however, only one of the two switches will dim the light. If you want both switches to dim the light, you must install two dimmers.

A single-pole dimmer switch is connected just like a standard single-pole switch. Some dimmers come with short wires instead of terminals. These are connected to the existing wires (black to black, white to white) with wire nuts.

When installing a dimmer on a three-way switch circuit, first mark the black common wire with a piece of tape. Connect it to the black lead from the switch with a wire nut. Then connect the other two leads with wire nuts, one to the white grounding wire and one to the red wire.

Ground Fault Circuit Interrupters

The chance of getting a serious shock from a short circuit in an appliance or lamp is reduced if the device is properly grounded through a three-prong plug. However, you might still get a shock if you are standing on a wet area, such as the bathroom floor. For this reason ground fault circuit interrupters (GFCIs) are now required in all bathroom, garage, and outdoor receptacles. They should also be used in kitchen, laundry, and workshop circuits. If there is a current leak anywhere in the circuit, the GFCI will shut off power within $\frac{1}{40}$ second.

The most commonly used GFCI is built into a receptacle. Used as the first receptacle on a given circuit, it will protect all the others. Another type of GFCI is installed directly in the breaker panel to protect the entire circuit. A third type can be plugged into an existing receptacle for temporary use.

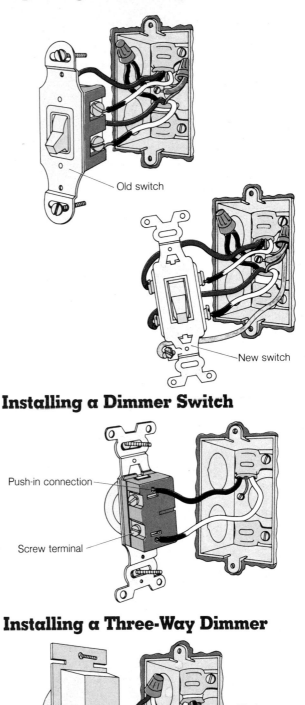

Replacing a Four-Way Switch

Old switch

New switch

Installing a Dimmer Switch

Push-in connection

Screw terminal

Installing a Three-Way Dimmer

Mark common wire with tape

MAINTENANCE, REPAIR & RESTORATION

Whether you own or rent, your home requires constant cleaning, periodic maintenance, and occasional repairs. It's human nature to put off the cleaning, defer the maintenance, and ignore the need for repairs whenever possible. Only major problems, such as a range that stops working, get immediate attention. Dust, frayed cords, faulty plugs, and lamps that flicker are minor problems often left to another day. In this chapter you'll discover that there are good, sound reasons for cleaning, maintaining, troubleshooting, and repairing your lighting.

Keep bulbs and fixtures clean and in good working order to show off your home to its best advantage.

BASIC MAINTENANCE

Your normal household maintenance routine should include attention to fixtures, switches, and plugs. Regular cleaning and inspection will ensure that fixtures function safely and efficiently.

Keeping Fixtures Clean

Cleaning light fixtures should be a part of the regular household routine. It maintains their appearance, prevents the metal parts from deteriorating, and keeps light output at a maximum. Pendants, wall lights, and uplights collect dust and dead insects. The residues from tobacco smoke quickly form a dull, sticky coating on fixtures and shades. The heat of the bulb also causes shades to attract dust.

Don't forget about safety when you clean a fixture. Be sure that the light is off; before removing the fixture for cleaning, cut the power by pulling the fuse or tripping the breaker for that circuit. A fixture that has been washed should be thoroughly dry before it is reinstalled.

Clean aluminum reflectors with a damp sponge or cloth and a mild detergent solution. If the reflector is stained, remove the stain with a small amount of metal polish first. Rinse off the detergent solution with clear water and dry the surface of the reflector with a soft cloth.

Glass cleaner can be used on glass surfaces, but don't use it on acrylic or other plastics. It can dull and damage them. Acrylic and plastic surfaces should be cleaned with warm soapy water and a soft cloth.

Use paper towels to blot the surface dry. Rubbing plastics dry with a cloth causes static electricity that attracts dust. (The tendency of plastic to attract dust can be frustrating, but don't use a household antistatic spray; it can damage the surface. Try going over the surface once more with a paper towel dampened in detergent and water and blot dry.)

Clean lampshades frequently to prevent dust buildup, which can pose a major cleaning problem, especially with light-colored shades. When cleaning a lampshade, remove it from the fixture. Gently brush fabric shades to remove dust; water may stain or discolor the fabric. Use a feather duster on paper shades. Then go over the shade lightly with an appropriate vacuum cleaner attachment.

Replacing Wire Plugs

A plug that is cracked or that has loose or broken prongs must be replaced. A plug that feels warm in use should also be replaced. If the plug doesn't fit tightly in the outlet, bend the prongs a little farther apart before deciding to replace it. Always replace a plug with one of the same kind.

The screw terminal plug is probably the most common. To replace it, first pry off the cardboard insulator at the base of

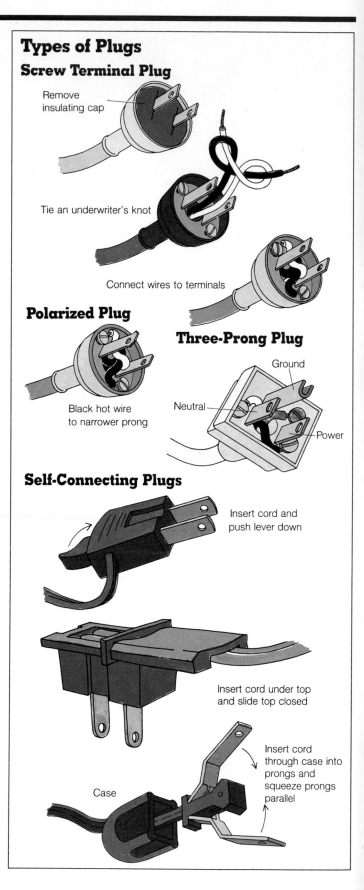

Types of Plugs

Screw Terminal Plug

Remove insulating cap

Tie an underwriter's knot

Connect wires to terminals

Polarized Plug

Black hot wire to narrower prong

Three-Prong Plug

Ground

Neutral

Power

Self-Connecting Plugs

Insert cord and push lever down

Insert cord under top and slide top closed

Case

Insert cord through case into prongs and squeeze prongs parallel

the prongs. Loosen the two screws holding the wires. Don't try to remove the screws completely, since they stop automatically at a certain point. Remove the wires from under the screws, untie the knot, and take off the old plug. If the cord has been damaged near the plug, cut it off behind the damaged section. Separate the wires for about 2 inches. If there is an outer layer of insulation, remove that first by cutting around it with a knife. Slip the wires through the new plug. Remove ½ inch to ¾ inch of insulation from the end of each wire. Tie an underwriter's knot to take up any strain on the wires if the plug is pulled out of its socket by the cord. Twist the stranded wire on each end clockwise. Then wrap each end clockwise under a screw and tighten it down. Unless the plug is a polarized type, it doesn't normally matter which wire goes to which screw.

A polarized plug is replaced in the same way as a screw terminal plug with one exception. One prong is always narrower on a polarized plug, and the hot wire must be attached to the narrower prong. Some appliances, such as TVs, are wired in such a way that the prongs must match the incoming power. The narrow prong is hot and the wide prong is neutral.

A three-prong plug is wired in much the same way as a standard screw terminal plug, but the black hot wire must be attached to the brass colored terminal screw and the white wire to the silver colored terminal screw. The bare grounding wire is attached to the green grounding screw terminal. In many cases the plug is encased

in the same insulation as the cord. If this type of plug is damaged, cut the cord just behind the plug, throw the old plug away, and replace it with a standard three-prong screw terminal plug.

Self-connecting plugs are used only with zip cord. They have small teeth that bite into the wires to provide contact. This eliminates the need to remove the insulation. Do not separate the wires, but make sure that the ends are cut square and evenly.

One common design has an arm that lifts up to expose a slot for installing the cord. Slit the end of the cord about ¼ inch and push it into the slot as far as possible. One side of the split wire will slide down into the plug; the other side will stay up. Press the arm back down into place to lock the plug securely onto the cord.

Another type of self-connecting plug has a slide-on top. To install it, slide the top off and place the zip cord on the open plug. Slide the top back, and the two small prongs inside the plug will pierce the wires and make contact.

The end-loading design has a pull-out prong section. Pull it out, insert the wire through the hole in the outer cover of the plug and push it into the prong section as far as it will go. Squeeze the prongs firmly and slip the cover over the unit.

Replacing In-Line Switches

A switch that is cracked, feels warm in use, gives off sparks, or makes a loud noise when operated is unsafe and must be replaced. There are two common types of in-line switches.

A *rocker switch* contains two screw terminals and an on-off rocker that opens and closes the circuit between them. To replace it, remove the screws and nuts holding the switch case together. Loosen the two screws holding the wire ends. Remove the wire ends from under the screws and discard the old switch. Separate the new switch and loosen the terminal screws. Reform the loop on each end of the cut wire and fit the uncut wire into the slot provided in the switch case. Wrap the wire loops around the screw terminals and tighten the screws. Fit the halves of the case together and install the screws and nuts securely. Plug in the cord and test the switch.

When installing this type of switch on a new length of cord, determine where it should go and carefully slit the insulation down the line between the molded wires for about 2 inches. Cut the smooth (hot) wire at the midpoint of the slit and remove about ½ inch of insulation from each cut end. Twist the stranded wire on each end clockwise and form a wire loop. Fit the loops around the screw terminals and tighten. Then reassemble the case and test the switch.

A thumb wheel switch contains pin contacts instead of screw terminals. To replace it, remove the screw and nut holding the case together. Separate the two halves of the case and pull the old switch off the cord. Separate the new switch and fit the uncut wire into the slot provided in the switch case. Assemble the case and apply sufficient finger pressure to force the pin contacts into the wire insulation. Install and

tighten the screw and nut. Plug in the cord and test the switch. To install a thumb wheel switch on a new cord, slit the insulation and cut the smooth (hot) wire at the midpoint, as for a rocker switch, but do not strip the ends of the wire.

Incandescent Lamps

Lamps with incandescent bulbs are all very much alike. They have only a few basic parts: the bulb, the socket, the switch, the cord, and the plug. These are the only components that can wear out, and each one can be easily fixed.

Understanding how the lamp is assembled not only allows you to make any necessary repairs, but also to make your own lamps from old bottles, pieces of driftwood, or anything. All the parts to assemble a lamp from scratch can be found in kit form in hardware or electrical stores.

To rewire an incandescent lamp, follow these steps.

1. Unplug the lamp and remove the bulb.
2. Unscrew the finial at the top of the harp and remove the shade. If the lamp has a two-piece harp, slide up the two metal sleeves at the base of the harp, squeeze the harp at the base, and lift it out.
3. The socket consists of four pieces: the outer shell, the insulating sleeve, the socket itself with its switch and terminal screws, and the socket cap. To separate the outer shell from the cap, depress the shell at the spot marked *press* and lift it up. If the shell will not come off, gently pry it free with a screwdriver. Remove the

cardboard insulating sleeve. Loosen the two terminal screws on the socket and remove the wires. Lift the socket from its cap.

4. Untie the underwriter's knot in the cord. Loosen the set screw at the base of the cap. Unscrew the cap from the threaded nipple. If there is no set screw and the cap does not unscrew easily, give it a hard turn.

5. Before removing the cord from the lamp pipe, tape the end of the new cord to the old one. As the old cord is pulled out of the lamp, it will pull the new one in. Untape the old cord and discard it.

6. Separate the first 2 inches of the new cord and tie an underwriter's knot at the top of the separation.

7. Strip off ½ inch to ¾ inch of insulation at the end of the wire and twist the strands clockwise. Loop the hot wire clockwise around the brass screw and the neutral wire clockwise around the silver screw. Tighten the screws.

8. Reassemble the lamp. Install a plug on the end of the cord.

Fluorescent Fixtures

Efficient as they are, fluorescent lights can go haywire. But if you understand how all the components work, fixing a fluorescent lamp is not much more difficult than repairing an incandescent one.

Tubes

Fluorescent tubes can be temperamental. Dirt or corrosion on the pins can cause problems; cleaning the pins can restore the tube to good working order. The life of a tube can sometimes be extended simply by removing it, turning it end for end, and reinstalling it in the fixture.

To remove a double-pin fluorescent tube, twist it a quarter turn in either direction and gently pull it out. To install a new tube, slip the pins into the slots and give the tube a quarter turn in either direction. On a single-pin model, push the tube back against the spring-loaded socket at one end until the pin at the other end is free and then lift the tube out. To install the new tube, put the pin on one end into the spring-loaded socket and push back until you can slip the pin on the other end into the other socket.

Starters

The starter, which is usually a small aluminum cylinder, is either built into the ballast or

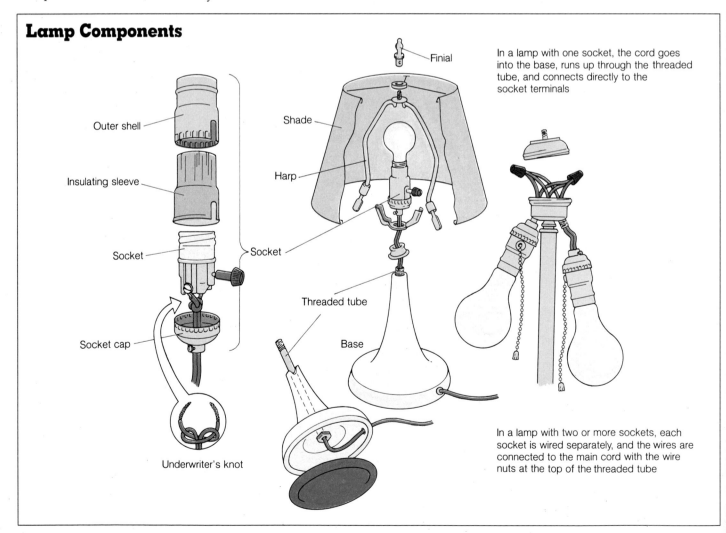

Lamp Components

Outer shell

Insulating sleeve

Socket

Socket cap

Underwriter's knot

Shade

Harp

Socket

Threaded tube

Base

Finial

In a lamp with one socket, the cord goes into the base, runs up through the threaded tube, and connects directly to the socket terminals

In a lamp with two or more sockets, each socket is wired separately, and the wires are connected to the main cord with the wire nuts at the top of the threaded tube

wired separately. Starters fail just about as often as tubes. Starter trouble is indicated by tubes that flicker constantly or glow only at the ends. Try re-seating the starter to correct a flickering tube; if this doesn't work, replace the starter. If only the ends of the tube light up, the starter must be replaced.

If the starter is in the ballast, remove it by giving it a quarter turn counterclockwise and then pulling it out. If the starter is separate from the ballast, shut off power to the lamp or un-plug it and then remove the wire nuts connecting the starter to the socket. In either case reverse the process to in-stall a new starter.

Sockets

Occasionally the sockets on a fluorescent lamp will be dam-aged or corroded beyond re-pair. To replace them, shut off power to the lamp and then remove the wires from each socket. If the wires are held by terminal screws, simply un-screw them. With push-in ter-minals the wires can be freed by pushing a small screwdriver into the slot next to the wires. On some models the sockets are held in place by a mounting screw; on other models they are slipped into position over the ends of the bracket. In the latter case remove the end cover plate and slide the socket out. Take the old sockets with you to the hardware store or the lighting shop so that you can buy exact replacements.

Ballasts

The ballast has a long life, but it's over when the fixture starts to emit a buzzing sound, a strong asphalt odor, or a black,

oily drip. On a fluorescent table lamp, the ballast is in the base. On a wall- or ceiling-mounted lamp, the ballast is a boxlike device behind the cover plate.

To replace it, unplug the lamp and then remove the flu-orescent tubes and the cover plate. Under the cover plate there will be up to eight color-coded wires. The black and white wires are connected with wire nuts to the corresponding wires from the power source. The power source grounding wire is usually attached directly to the metal channel with a sheet-metal screw. The remain-ing wires go to the lamp sock-ets. Make a note or draw a diagram of the wire connec-tions. Cut the wires into the ballast and free it by removing the mounting screws.

Take the old ballast to the lighting shop and buy a re-placement with the same volt-age rating and design. Each brand of fixture has its own specific wiring. Follow the in-stallation procedure and the wiring diagram supplied with the new ballast when you in-stall it.

Troubleshooting

Many problems are best reme-died by a professional electri-cian, but the homeowner can solve a few of them. Remember to work safely. Always turn off the power before starting to work on a circuit, and make sure that it is off by checking with a voltage tester. Do not work on a wet or damp surface; lay down a rubber mat or dry boards to stand on if necessary. Make sure that your hands are dry before you touch the wiring. This is especially

important when testing a live circuit with a voltage tester. When using the voltage tester to check a connection, handle it only by the insulation, never

by the bare probes. Don't touch any metal while you are testing the circuit. Refer to the safety rules in the previous chapter before starting any repair.

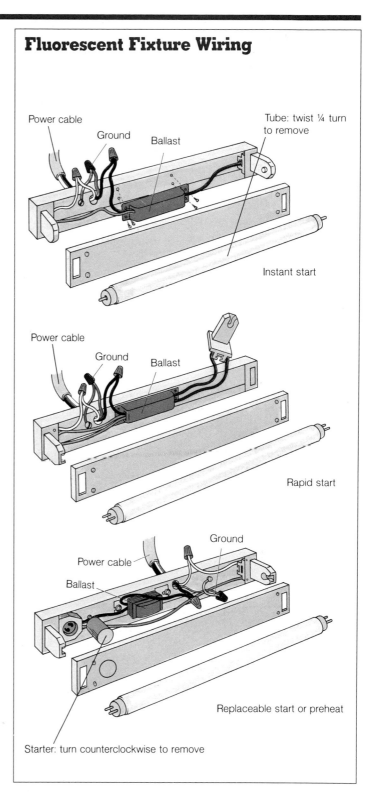

Fluorescent Fixture Wiring

Power cable
Ground
Ballast
Tube: twist ¼ turn to remove
Instant start

Power cable
Ground
Ballast
Rapid start

Power cable
Ground
Ballast
Replaceable start or preheat
Starter: turn counterclockwise to remove

DECORATIVE TOUCHES AND RESTORATIONS

Table lamps are usually chosen partly for their visual appeal, and a well-chosen finial can enhance that appeal. Canopies can do the same for ceiling fixtures. An antique lamp, restored to working order, can be especially attractive.

Harps, Finials, and Canopies

A *harp* positions the shade on a table lamp. A detachable harp consists of a wing, which fits over the lamp pipe and is retained by the socket, and the harp proper, which locks into each end of the wing with small metal sleeves. A screw-on harp can be used to adapt a lamp designed for a clip-on shade to accept a regular shade and finial.

Both types of harp have a swiveling shade holder at the top. The threaded stud on the swivel accepts the finial. The swivel makes it possible to adjust the position of the shade to any angle.

A *finial* holds the shade in place on the lamp harp. It also provides a decorative touch. On many lamps the finial is nothing more than a knurled nut that screws onto the shade holder. However, finials are available in many sizes and designs, from simple turnings to pineapples, eagles, and so forth. For elaborate lamps there are finials made of delicate filigree, crystal, and even a variety of precious stones.

A *canopy* is the cover used with ceiling fixtures to hide the wires and the ceiling box from view. Canopies vary in design from plain metal caps to ornate sculpted medallions. Canopies are attached in various ways. The simplest and most common type has a center hole for the cord to pass through.

Screws inserted through mounting holes near the edge of the canopy hold it to a crossbar fastened to the ceiling box.

Canopies may be functional as well as decorative. One design has a hook in the center to hang a chain lamp or a chandelier and a built-in outlet to plug in the fixture. This makes it easy to remove the lamp or chandelier for cleaning or to replace a bulb. Other canopies are designed in such a way that it will not be necessary to repair the ceiling when the fixture is permanently moved. Others still are designed to cover unwanted recessed lighting.

Electrifying a Glass Kerosene Lamp

Lamps originally designed for burning kerosene are still plentiful today. They can be found in all shapes, sizes, and styles, from expensive cased glass and cut glass to inexpensive plain molded-glass designs. If you have an original kerosene lamp, fine. If you're thinking of buying one to make it into a lamp, beware. The market is filled with modern reproductions of these old lamps. If you buy a reproduction thinking it's an original and then discover your mistake, you may have second thoughts about converting it.

Converting a kerosene lamp to electricity is simple; making it useful, however, takes a little more work. To convert it, just buy a prewired socket burner to replace the old wick burner. Remove the chimney and unscrew the wick burner from the base. Screw the prewired socket burner in its place. Add a bulb, install the chimney, and you have an electrified kerosene lamp ready to go.

To make it useful as well as ornamental, some sort of shade should be added to shield the glare of the bulb. It is possible to use a shade that fits down over the chimney, but keeping the chimney limits the size of the bulb that can be used. Never use a bulb of more than 75 watts inside a chimney. The heat may crack the glass.

To use a more powerful bulb, wire the lamp with a harp and use a standard shade. (Standard shades also come in a wider variety of designs.) For this conversion, get a brass oil lamp adapter instead of a prewired socket burner. You will also need a suitable threaded nipple, a neck, a round brass locknut, a side-hole socket, a harp, a finial, and a cord with a wall plug. Adapters, nipples, and necks come in different sizes; be sure to get the right size to fit your particular lamp.

To wire a kerosene lamp:

1. Screw the adapter into the lamp neck.
2. Screw the nipple into the adapter.
3. Add the neck and secure it in place with the brass locknut.
4. Install the harp wing on the nipple. Then screw the socket cap into place.
5. Connect the lamp cord to the socket terminals and seat the socket in the cap.
6. Slip the cardboard insulator over the socket.
7. Seat the outer metal shell in the socket cap and push it down until it clicks into place.
8. Fit the harp into the harp wings.
9. Install the shade and attach the finial.

Restoring Antique Lamps

Whether you want to install authentic fixtures in an old house or put an old fixture into safe working order, restoration can be time-consuming and expensive. Genuine antique lamps are often incomplete. Although electrical parts, such as sockets and cords, are easily obtained, you may have to settle for reproduction fixture components.

Reproduction fixtures as well as parts are handmade in small shops and are sold through catalogs. Some of them are faithful copies of old designs. Others are adaptations of old designs, and others still are simply meant to evoke the feel of a particular period. Unless you know a good deal about antique lighting, or can find someone who does, it is not easy to tell the difference between a good reproduction and the genuine article.

Fixtures that predate the era of electricity will have to be electrified. This should be done carefully to avoid drilling holes or removing original components (which will destroy the value of the fixture). Any work done should be reversible; that is, it should be possible to reconvert the fixture to its original state.

Here are a few useful hints for restoring old light fixtures.

• To clean and polish old brass parts, disassemble them and soak them in a half-and-half solution of ammonia and water to remove the old lacquer. Use extrafine steel wool to remove stubborn spots. Finish by applying a good-quality brass polish and buffing.

• To refurbish wood parts, first clean the finish. Then, using extrafine sandpaper, sand off the imperfections accumulated in the top surface over the years with the grain until the parts are satin-smooth.

• Cover metal parts, such as locknuts, with a piece of cloth before tightening them with pliers. This will prevent damage to the milling or other ornamentation.

• Remove all wiring from the fixture and replace it with new wires. Do this even if the old wires seem to be in good condition.

• When redoing a wall sconce from the gaslight period, use a headless brass reducer to adapt the gas pipe to a lamp pipe. Attach the lamp pipe to the wall outlet.

• The hollow arms of a single-light hanging fixture originally wired with rayon-covered wire may be too small to accept zip cord. If they are, carefully split the zip cord into separate wires and run them to the socket through two different arms.

Troubleshooting Plug-in Lights

If an incandescent bulb in a plug-in light flickers or does not come on at all, run through this quick checklist to isolate the problem.

1. Remove the bulb and insert it into a lamp that you know is working. Check if it lights.
2. If the bulb lights, plug the lamp into another outlet. See whether it works. If it does, the problem is in the receptacle.
3. If the lamp does not work, inspect the plug for cracks or loose wires. Make sure that the prongs fit tightly in the receptacle. Spread them slightly to achieve a tight fit. If the plug is cracked, replace it with a new one.
4. Make sure that the cord is in good condition and that the wires are not disconnected. Tighten the terminal screws on the plug. Self-connecting plugs can lose contact with the zip cord and may need to be removed and reinstalled.
5. Make sure that the spring terminal at the socket base bends upward enough to make firm contact with the bulb. If it does not, pry it up slightly with your finger or a screwdriver.
6. Disassemble the lamp and check the socket and switch with a continuity tester. Attach the alligator clip to the metal socket and touch the probe to the silver colored terminal screw. Replace the socket if the tester does not light. To test the switch, connect the alligator clip to the brass colored screw and touch the probe to the spring terminal at the base of the socket. If the tester does not light when the switch is turned on, replace the socket and switch assembly.

Table lamps and accent lights in the cabinet provide background lighting for this room, while track lights give the necessary task lighting and general illumination.

GLOSSARY OF LIGHTING TERMS

You will encounter the following terms both in this book and in designing and working on a lighting system. Familiarize yourself with them now and refer to the glossary as necessary.

Absorption A measure of the amount of light that is absorbed by rather than reflected from an object. Surfaces that are black or dark colored or are heavily textured absorb the most light. *See also* Reflectance.

A-bulb A type of incandescent bulb; the common household light bulb.

Alternating current (AC) An electric current that reverses its direction at regularly recurring intervals.

Amp An abbreviation for ampere.

Ampacity How much current, expressed in amperes, a conductor can carry.

Ampere The SI unit that measures the flow of electrical current. *See also* SI.

Apparent brightness (luminosity) A subjective impression of how bright an area appears.

Approved An endorsement stating that minimum standards established by an authority have been met.

Arc, electric A visible, sustained discharge of electricity that bridges a gap in a circuit or between electrodes.

Armored cable (BX) A flexible, metal-sheathed cable used for indoor wiring.

Background lighting Subdued general lighting, indirect and unobtrusive.

Baffle A device used with a light fixture to prevent glare.

Ballast A magnetic coil that adjusts current through a fluorescent tube, providing the current surge to start the lamp.

Bayonet base A lamp base with two lugs to hold a tungsten bulb in a lamp socket.

Beam The spread of direct illumination from a spotlight or floodlight reflector bulb. Light intensity at the edges is half that at the center.

Box *See* Outlet box.

Branch circuit A circuit that supplies a number of outlets for lights or appliances.

Bulb A general term for an electric lamp, especially an incandescent lamp.

BX *See* Armored cable (BX).

Cable A conductor consisting of two or more wires that are grouped together in a protective outer covering.

Candela The SI unit that measures light intensity. *See also* SI.

Candlepower The intensity of a light source, measured in candelas.

Circuit The path of electrical current along the supply cables to the light fixtures and receptacles and back to the source.

Circuit breaker A switch in an electrical service panel that opens the circuit when the current in the circuit exceeds a predetermined amount; can be reset.

Code, National Electric (NEC) A set of rules sponsored by the National Fire Protection Association, under the auspices of the American National Standards Institute, to protect against electrical hazards.

Color appearance The apparent color of light given off by a particular source.

Color coding The identification of conductors by color.

Color rendition The effect of a particular light source on the visual appearance of a colored surface. A lamp may have a color rendering index (CRI) between 1 and 100; the larger the number, the truer the color.

Color temperature A measure of the redness or blueness of a light, measured in degrees Kelvin (°K).

Conductor A low-resistance material, such as copper wire, through which electricity flows easily.

Conductor, bare A conductor that has no covering or insulation.

Conduit A metal or fiber pipe or tube that is used to enclose electrical conductors.

Continuity An uninterrupted electrical path.

Contrast The subjective difference in the brightness of the lighting between two parts of a visual field when seen successively or simultaneously.

Cool-beam bulb A bulb with a dichroic reflector that reflects visible light but transmits infrared radiation, reducing heat in the beam.

Cord A flexible length of insulated wire that connects a light fixture to a power source; plugs into a receptacle.

Current The flow of electricity through a conductor, such as a wire, measured in amperes.

Cycle A complete positive and negative alternation of a current or voltage.

Device A unit or component that carries, but does not use, current, such as a junction box or a switch.

Diffused lighting Lighting filtered evenly through diffusing or translucent materials to prevent glare; used especially when fixtures must be located in the line of sight.

Diffusion   The scattering of the light emitted by a lamp, whether by a diffusing coating on the bulb, by a lens or grill on the fixture, or by other means. Diffused light is softer, creates less glare, and produces soft shadows.

Dimmer   A switch that dims the brightness of lighting fixtures by reducing the flow of current.

Direct current (DC)   An electrical current that flows in only one direction.

Directional lighting   Light designed to illuminate a surface, object, or task from a particular direction.

Direct lighting   Light produced directly from a fixture without being reflected from surfaces.

Downlight   A ceiling or track-mounted fixture that directs light downward.

ER bulb   An incandescent bulb with an internally silvered surface, delivering a light spread between a spotlight and a floodlight.

Energy efficiency   In lighting, the amount of light (measured in lumens) produced by a bulb for each watt of electricity consumed.

Filament   The thin tungsten wire in a light bulb that emits light when heated to incandescence by an electrical current.

Fish tape   Flat steel spring or nylon wire with hooked ends; used to pull wires through conduits or walls.

Fitting   An accessory (such as a bushing or a locknut) used on wiring systems to perform a mechanical, not electrical, function. (Also a term for a lighting fixture in some countries, such as Canada and England.)

Fixture   *See* Light fixture.

Flexible cable (flexible cord)   A conductor that is made of several strands of small-diameter wire.

Floodlight   A tungsten lamp that produces a bright, broad beam of light.

Fluorescent bulb   A bulb in which a coating on the inside of a glass tube is made to glow by an electrical current.

Footcandle   A unit that measures light; the amount of illumination when 1 lumen falls on 1 square foot of surface.

Footlambert   The flow of light from a uniform diffuser emitting 1 lumen per square foot.

Framing projector   A spotlight using attachments that give control over the shape and focus of the beam; especially useful for picture lighting.

Fuse   A safety device containing a band of metal that melts to interrupt a circuit when the current in the circuit exceeds a predetermined safe level for a specific time period.

Glare   Distractingly bright light that interferes with vision.

Grazing light   The positioning of a light source to bring out the texture or surface dimension of a wall, door, or other element.

Ground   A connection between an electrical circuit and the earth or a body serving in place of the earth; a grounding wire is always white or bare.

Ground fault circuit interrupter (GFCI)   A safety device that interrupts a circuit in about $1/40$ second if it detects any leakage of current, to prevent danger of shock.

Grounding wire   In a cable, the wire (usually green) that carries no current; used as a safety measure.

Halogen bulb (tungsten-halogen or quartz incandescent bulb)   An incandescent bulb containing a tungsten filament within a pressurized fused-quartz bulb that is filled with a halogen gas; emits a bright, yellow-white light.

Hanger   A metal or insulated strap used at intervals to support electrical cable between points of connection.

High-intensity discharge bulb (HID bulb)   A bulb that produces light when electricity excites specific gases within a pressurized container. Includes mercury vapor, metal halide, and high-pressure sodium bulbs; requires special fixtures and ballasts.

Hot wires   The wires of a house circuit that are not connected to a ground and that carry power to outlets and appliances. May be any color except white and green.

Incandescent bulb   A bulb that produces light when electricity heats a metal filament to incandescence; a standard household lamp emitting a yellow-white light.

Indirect lighting   Lighting that is reflected from a ceiling or one or more walls.

Insulation   Materials that do not carry current; used on the outside of wires and in the construction of electrical devices.

Insulator   A nonconductor that is used to support a conductor that carries current.

Junction box   A box in which several conductors (wires) are joined together.

Knockout   A circular die-cut impression in outlet boxes that has not been completely severed and that must be removed to accommodate wiring.

Lamp   The technical name for what is commonly called a light bulb. It is a tube, usually of glass, in which a filament, gas, or coating is excited by electricity to produce light.

Light bulb   *See* Lamp.

Light fixture   The housing for a bulb and socket. It usually contains a reflector and electrical wiring connected to a power source. It may also contain a lens to protect the bulb, and a ballast if it is intended for use with fluorescent or HID bulb. It may direct or control the flow of light.

Light output   The amount of light emitted by a lamp, measured in lumens.

Light source   The combination of a bulb and a fixture; used to illuminate an environment.

Live wire   A wire that carries current.

Louvers   A set of parallel metal, plastic, or wood slats installed on a light fitting to prevent direct viewing of a light source from certain angles; eliminates glare.

Low-voltage lighting   Lighting that operates on 12-volt current rather than on the standard 120 volts (a few systems use 24 volts). Power is supplied by a transformer, which itself is connected to 120-volt current.

Lumen   A unit that measures the amount of light emitted by a lamp. One lumen is the amount of light emitted by one standard candle at a distance of 1 foot. Lumens of emitted light are measured at the light source.

Luminosity   *See* Apparent brightness (luminosity).

Meter, electric   A device that measures how much electricity is used.

National Electric Code   *See* Code, National Electric (NEC).

Neon bulb   A bulb containing neon (an inert gas) at low pressure; emits a yellowish red glow when voltage is applied.

Neutral wire   All current that flows through the hot wire and is not used by lights and appliances must flow through the neutral wire back to the ground terminals.

Nonconductors   Materials, such as glass, porcelain, and rubber, through which electric current does not flow.

Nonmetallic cable   Non-metal-sheathed electrical cable used for indoor wiring.

Open circuit   An electrical circuit with a physical break in the path through which no current can flow; caused by opening a switch, disconnecting a wire, or burning out a fuse, for example.

Outlet box   A metal or plastic box in which electrical wiring is connected to electrical components.

Overload   Current demand exceeding that for which the circuit or equipment was designed; usually causes the fuse or circuit breaker controlling the circuit to blow.

PAR bulb   A sealed-beam incandescent bulb with a front of heat-resistant glass, internally aluminized to reflect and spread a powerful beam of light.

Pendant   A light fixture that hangs from the ceiling.

Plug (attachment)   A device that is inserted into a receptacle to establish an electrical connection to a circuit. *See also* Receptacle.

Polarized plug   A plug with blades designed to enter a receptacle in only one orientation.

Polarizing   The use of color to identify wires throughout a system to make sure that hot wires will be connected only to hot wires, and that neutral wires will run back to the ground terminals in continuous circuits.

Power   The rate at which work is being done. Electrical power is measured in watts. *See also* Watt.

Raceway   A channel that supports electrical wires or cables.

Rated life   The number of hours at which 50 percent of lamps will fail under standard test conditions.

Receptacle   A contact device that is installed at the outlet to supply current to a single plug.

Receptacle outlet   An outlet where one or more receptacles are installed.

Reflectance   The ratio of the lumens reflected from a surface to the lumens falling on it. Reflectance is highest from objects or surfaces that are smooth and light colored.

Reflector   A mirrored or polished surface designed to project the beam from a light source in a given direction. Can be a part of the bulb or part of the fixture.

Reflector bulb (R bulb)   An incandescent bulb with an internally silvered surface, used for spotlighting or floodlighting.

Screw terminal   A means of connecting wiring to devices, using a threaded screw.

Service panel   The main panel through which electricity is brought into a building and distributed to the branch circuits.

Shade   A device used on a light fixture; prevents glare, controls light distribution, diffuses and may color the light emitted.

Short circuit   An improper connection between hot wires, or between a hot wire and a neutral wire.

SI   An internationally agreed-upon system of scientific measuring units.

Solderless connector (wire nut)   A mechanical, typically plastic-insulated, device that can be fastened over the exposed and joined ends of several wires to make a firm connection.

Spill light   Light that falls outside the main profile of a beam.

Splice   A connection that is made by joining two or more wires.

Split receptacle   A dual receptacle in which each of the two units is connected to a different branch circuit rather than to a common circuit.

Spotlight   An incandescent bulb producing a bright, highly focused beam of light.

Starter   A device used with many fluorescent bulbs to strike an arc between the electrodes when the fixture is turned on.

Stranded wire (stranded cable) A quantity of small conductor wires that are twisted together to form a single conductor.

Switch A device that is used to connect and disconnect the flow of current or to divert current from one circuit to another; used only in hot wires, never in grounding wires.

Task lighting Localized lighting used to illuminate a particular activity.

T-branch A right-angle extension to a circuit to provide electricity to an additional fixture, switch, or receptacle.

Three-way switch A type of switch, two of which can be used to control a light from two different directions.

Transformer A device that converts current of one voltage into current of another voltage; an essential component of any low-voltage lighting system.

UL label A label applied to manufactured devices that have been tested for safety by Underwriters' Laboratories, Inc. and approved for placement on the market. These laboratories are supported by insurance companies, manufacturers, and other parties interested in electrical safety.

Underwriter's knot A knot that is used to tie two insulated conductors at the terminals inside an electric plug; used to relieve strain on the terminal connection.

Uplight A fixture that directs light upward to the upper walls or ceiling of a room; illuminates by reflection.

Uplighting The lighting of an object or surface from below.

Volt A unit that measures electrical pressure; comparable to pounds of pressure in a water system.

Voltage The pressure difference that causes electric current to flow.

Voltage drop A loss of electrical current caused by overloading wires or by using excessive lengths of undersize wire. Frequently indicated by the dimming of lights at the ends of wire runs.

Watt A unit that measures electrical power. Volts × amperes = watts of electrical energy used. The unit by which electric companies meter power supplied.

Wire An electrical conductor in the form of a slender rod.

Wire gauge A standard numerical method of specifying the physical size of a conductor. The American Wire Gauge (AWG) system is the most common.

Wire nut A device that uses mechanical pressure, rather than solder, to establish a connection between two or more connectors.

Wiring diagram A drawing, in symbolic form, showing conductors, devices, and connections.

Supply Sources

Many electrical supplies and fixture components can be bought from local hardware stores, lighting shops, distributors, or manufacturers. Try the following sources, grouped by category, when you cannot find what you need locally.

Bulbs
Common and not-so-common decorative and standard bulbs

Light Bulbs Unlimited
14456 Ventura Blvd.
Sherman Oaks, CA 91423

Fixture Components (Nonelectrical)
Harps, finials, loops, chains, risers, lamp kits, decorative hooks, shade fittings, etc.

Aetna Pipe Products
  Company of Illinois
3515 West Armitage Avenue
Chicago, IL 60647

Allen-Stevens Conduit
  Fittings Corp.
29 Park Avenue
Manhasset, NY 11030

Cable Electric Products, Inc.
Providence, RI 02907

Chilo Manufacturing and
  Plating Co.
206-10 South Kedzie Avenue
Chicago, IL 60623

Freeman Products
86 State Highway 4
Englewood, NJ 07631

Kirks Lane Lamp Parts
1445 Ford Road, Box 519
Cornwells Heights, PA 19020

Lite House
369 West Portal Avenue
San Francisco, CA 94127

Span-O-Chain, Ltd.
7412 Bergenline Avenue
North Bergen, NJ 07047

Period and Reproduction Fixtures
Authentic Designs
The Mill Road
West Rupert, VT 05776

Brass Light Gallery
719 South Fifth Street
Milwaukee, WI 53204

King's Chandelier Co.
Box 667
Eden, NC 27288

Lt. Moses Willard & Co.
1156 U.S. 50
Milford, OH 45150

Period Lighting Fixtures
1 Main Street
Chester, CT 06412

Rejuvenation House Parts
901 North Skidmore
Portland, OR 97217

Victorian Reproduction
  Lighting Co.
Box 579
Minneapolis, MN 55458

Restored Antique Fixtures
Authentic Lighting
558 Grand Avenue
Englewood, NJ 07631

Greg's Antique Lighting
12005 Wilshire Boulevard
Los Angeles, CA 90025

Illustrious Lighting
1925 Fillmore Street
San Francisco, CA 94115

Yankee Craftsman
357 Commonwealth Road
Wayland, MA 01778

Shades
Lampcrafters
97 Somerset Street
North Plainfield, NJ 07060

Lite and Shade, Inc.
1193 Lexington Avenue
New York, NY 10028

INDEX

Proof-of-Purchase
0-89721-227-4

U.S. / Metric Measure Conversion Chart

	Symbol	*Formulas for Exact Measures* When you know:	Multiply by:	To find:	*Rounded Measures for Quick Reference*		
Mass (Weight)	oz	ounces	28.35	grams	1 oz		= 30 g
	lb	pounds	0.45	kilograms	4 oz		= 115 g
	g	grams	0.035	ounces	8 oz		= 225 g
	kg	kilograms	2.2	pounds	16 oz	= 1 lb	= 450 g
					32 oz	= 2 lb	= 900 g
					36 oz	= 2¼ lb	= 1000 g (1 kg)
Volume	tsp	teaspoons	5.0	milliliters	¼ tsp	= 1/24 oz	= 1 ml
	tbsp	tablespoons	15.0	milliliters	½ tsp	= 1/12 oz	= 2 ml
	fl oz	fluid ounces	29.57	milliliters	1 tsp	= ⅙ oz	= 5 ml
	c	cups	0.24	liters	1 tbsp	= ½ oz	= 15 ml
	pt	pints	0.47	liters	1 c	= 8 oz	= 250 ml
	qt	quarts	0.95	liters	2 c (1 pt)	= 16 oz	= 500 ml
	gal	gallons	3.785	liters	4 c (1 qt)	= 32 oz	= 1 liter
	ml	milliliters	0.034	fluid ounces	4 qt (1 gal)	= 128 oz	= 3¾ liter
Length	in.	inches	2.54	centimeters	⅜ in.	= 1 cm	
	ft	feet	30.48	centimeters	1 in.	= 2.5 cm	
	yd	yards	0.9144	meters	2 in.	= 5 cm	
	mi	miles	1.609	kilometers	2½ in.	= 6.5 cm	
	km	kilometers	0.621	miles	12 in. (1 ft)	= 30 cm	
	m	meters	1.094	yards	1 yd	= 90 cm	
	cm	centimeters	0.39	inches	100 ft	= 30 m	
					1 mi	= 1.6 km	
Temperature	° F	Fahrenheit	⅝ (after subtracting 32)	Celsius	32° F	= 0° C	
					68°F	= 20°C	
	° C	Celsius	⅝ (then add 32)	Fahrenheit	212° F	= 100° C	
Area	$in.^2$	square inches	6.452	square centimeters	1 $in.^2$	= 6.5 cm^2	
	ft^2	square feet	929.0	square centimeters	1 ft^2	= 930 cm^2	
	yd^2	square yards	8361.0	square centimeters	1 yd^2	= 8360 cm^2	
	a.	acres	0.4047	hectares	1 a.	= 4050 m^2	